Contents

For Niall Ferguson

War and Its Causes

Jeremy Black
University of Exeter

ROWMAN & LITTLEFIELD
Lanham • Boulder • New York • London

Executive Editor: Susan McEachern
Editorial Assistant: Katelyn Turner
Senior Marketing Manager: Kim Lyons

Published by Rowman & Littlefield
An imprint of The Rowman & Littlefield Publishing Group, Inc.
4501 Forbes Boulevard, Suite 200, Lanham, Maryland 20706
https://rowman.com

6 Tinworth Street, London SE11 5AL, United Kingdom

British Library Cataloguing in Publication Information Available

Library of Congress Cataloging-in-Publication Data
Names: Black, Jeremy, 1955– author.
Title: War and its causes / Jeremy Black.
Description: Lanham : Rowman & Littlefield, [2019] | Includes bibliographical references and index.
Identifiers: LCCN 2018046561 (print) | LCCN 2018052520 (ebook) | ISBN 9781538117927 (ebook) | ISBN 9781538117903 (cloth : alk. paper) | ISBN 9781538117910 (pbk. : alk. paper)
Subjects: LCSH: War—Causes. | War—History. | Military history, Modern.
Classification: LCC U21.2 (ebook) | LCC U21.2 .B525 2019 (print) | DDC 355.02—dc23
LC record available at https://lccn.loc.gov/2018046561

Printed in the United States of America

Abbreviations

Add.   Additional Manuscripts
ADM   Admiralty Papers
AN   Paris, Archives Nationales
BL   London, British Library
BN   Bibliothèque Nationale
CAB   Cabinet Papers
DRO   Exeter, Devon Record Office
FO   Foreign Office Papers
NA   London, National Archives
NAF   Nouvelles Acquisitions Françaises
SP   State Papers

Preface

The causes of war are a crucial topic for the modern world, one that historical study is well placed to illuminate. Indeed, the past is the "data set"; the present is only a wisp between past and a future that is unknown.

The frequency of conflict in the modern world, and the extent of the reappraisal of earlier warfare, are the backgrounds to this book. I have written before on aspects of the question, notably in *Why Wars Happen* (1998), but not at length for more than two decades. Present concerns and developments throw light on the past. In particular, the salience of religion in politics and international relations from the 1990s throws light both on the past and on analyses produced in the 1970s and 1980s, especially those that underrate the role of ideology. So also with the role in the 2010s of authoritarian leaders keen to reject Western norms on international relations and war. Thus, analyses produced in a period that was fundamentally motivated by secular considerations and norms now need supplementing by others that pay more attention to religious ones and to the role of the individual search for *gloire*.

In rethinking my earlier work, I have been helped by both the range and quality of the scholarship of others and by the opportunity to lecture widely. I have particularly benefited from speaking at a 2018 conference on "Gaming the Future: Markets and Geopolitics in an Uncertain World," and attending another on "The Analogy Trap? Thucydides, History and Theory in the Age of Trump," and from speaking at the Naval War College, the National Defense University, the Royal College of Defence Studies, the Universities of North Carolina, Chapel Hill and Oxford, Assumption, High Point, Mary Washington, North Georgia, and Roger Williams universities, the Citadel, the New Jersey Institute of Technology, the New York Historical Society, the World Affairs Council of Delaware, the American Museum in Bath, Radley College, the Athenaeum, and the D Group.

I have also benefited greatly from the comments of Paola Bianchi, Mark Danley, Kelly DeVries, John France, Paul Gelpi, Bill Gibson, Ned Lebow, Luigi Loreto, Wayne Malbon, David Morgan-Owen, Kaushik Roy, Rick Schneid, Doug Stokes, and Heiko Werner Henning on an earlier draft, although all errors are my responsibility. Susan McEachern has proved a most supportive editor. It is a great pleasure to dedicate this book to Niall Ferguson, whom I first met more than thirty years ago, and whose books and other publications have consistently provided both insight and interest.

Chapter One

What Is War?

What is war? There is no agreed definition, legal or otherwise, today among the multitude offered, and none that works across time and cultures. And that is simply if we are thinking of conflict between groups of humans with some organization, no matter how sophisticated, and not that between humans and other species or those among the latter. The situation is complicated by the whole host of what could be seen as "non-traditional conflicts," whether the term is applied, and has long been applied, or not. Some of this host have overlapped with killings, others not. Indeed, the classic instance of the use of the term *war* across much of history is that of conflict over religion, both between religions and, more profoundly, between good and evil. In this definition, fighting is but part of a more existential conflict, with the latter greatly affecting understandings of war. In presentation and belief, this is total war, a term usually seen, somewhat misleadingly, as arising in the twentieth century to discuss the new reality of World War I,[1] but, in practice, a term better applied to the long-standing elemental struggle between religions. This struggle reflected the strong belief that there was one "true" religion.

A different and far more recent instance of a use of the term war that very much overlaps with violence is that of "global war on terrorism." With an overlap, but less of an overlap, there is "war on drugs" and "war on crime," let alone cancer, poverty, ignorance, et al. This flood of "wars" leads to a normalization of the term in English, a process recently seen also with other words in that language. The French do not use "*guerre*," the Germans "*krieg*," nor the Italians and Spaniards "*guerra*," in the same way that English speakers, more particularly Americans, overuse it.

As an instance of overlap, the scale of criminality can be such that the casualties surpass those in many individual conventional wars between nor-

1

mal states, or, indeed, rebellions or coups. Thus, in Mexico in 2017, there were over 23,000 murders. Criminal networks there have reached levels of organization close to a "normal" army. Law and order "conflicts" are scarcely new, with all the issues these pose for definition. For example, in nineteenth-century Australia, British settlers formed private militias to protect their property from bushrangers and cattle and sheep thieves.

There are many types of conflict that are rarely murderous, including class warfare, culture wars, the battle of the sexes, generational conflict, and history wars. That is not a complete list. Moreover, it can be expanded, but also complicated, if other languages and cultures are considered. In addition, any and every description and/or analysis of relationship as focused on power, confrontation, and force extends the extent to which the idea and language of war play a major role in linguistic usage. Indeed, the language of war has come to be applied to everything and anything held to require an effort. Similarly, strategy is now applied to everything requiring a plan, whether running a business or a life, "process management," organizing a party, or deciding how to spend the morning.

Lack of linguistic precision, however, helps keep words valid, or at least in use, for the particular contexts and meanings of the past may change. That remark may appear preposterous in the case of large-scale conflict, but war to be war no longer requires an origin in the sovereignty of the state, or a formal process of declaration, as in some definitions in the past. Moreover, war to be war arguably never did.

Whether or not current linguistic usage is appropriate, this situation means that the language of war is far from coterminous with warfare as the latter is generally understood. For the sake of clarity, it is appropriate to offer a more restricted working definition. To do so accepts the consequent difficulty of boundaries with "non-war," and the extent to which any definition inevitably raises questions of authorial subjectivity, and of cultural and chronological specificity: this is the definition offered by "a" in period "b." Nevertheless, the process is necessary in order to offer some clarity, a situation also seen with other words whose meaning has been infinitely extended, such as revolution and strategy.

In functional terms, war can be seen as organized large-scale violence, both organized (politically and/or militarily) and large-scale. This definition separates war from, say, the actions of an individual, however violent the means or consequences: one individual poisoning a water supply could kill more than died in the Anglo-Argentine Falklands War of 1982. This definition also separates war from nonviolent action, however much, as in "direct action," such action can be an aspect of coercion. The definition also opens a gap with large-scale violence in which the organization is not that of, or for, war, for example football hooliganism. Hooligan groups plan for "battle" with each other, but this is not war even if the hostility is long-standing. Each of these points and caveats can be detailed and qualified, but they draw attention not only to fundamental issues of definition, but also to real difference.

War can also be approached in cultural, social, and ideological terms, namely as the consequence of bellicosity. These aspects repay examination. They focus on the importance of arousing, channeling, and legitimating violent urges and of persuading people to fight and, crucially, kill and run the hazard of being killed. Particularly in battle, but also in other combat roles such as sieges, which, in fact, were more important in medieval Christendom, people need to agree to put themselves at risk, to resist attack, and to advance across the "killing ground." Others need to be confident that they will do so, and thus preserve armies from dissolution. Without this, there would be no war. Leaving aside slave armies, the willingness or, rather, a large number of "willingnesses," acceptance and a large number of acceptances, are crucial to the causes of war. They are a conflation of long-term anthropological and psychological characteristics with more specific societal and cultural situations.

Whether, and to what effect, these propensities to organized conflict have altered over time is a historical question, and one made more complex (and, paradoxically, simpler) by recognizing that there is no single situation at particular moments. Moreover, for most of history, the nature of the sources is, generally, at best suggestive as to motivation. In addition, detailed work on the pressures and goals affecting conduct is often limited. This ensures that it is still possible to debate the causes of many individual conflicts, as well as warfare in a particular period.

So also with the occasions of individual conflicts. In 2018, was the reimposition of US sanctions on Iran and the abandonment of the Iran nuclear deal an act of war, or just a hostile step? Was it a response to Iran's "bloody ambitions" as President Trump declared, and thus a form of defensive step? As a linked occurrence, what did the significance of the mobilization of the Israeli military signify, and notably in response to "irregular activity" by Iranian forces in Syria. This was followed by an Israeli air attack on a Syrian base, and, subsequently, by full-scale Israeli air attacks on Iranian positions in Syria. Was there cooperation between the United States and Israel and, if so, how far did the actions of the latter compromise the former? What is the moment of war, and, therefore, its content?

The approach to discussing war with which we are most familiar is that of organized conflict between sovereign states, with the conflict begun by a specific act of policy at a particular moment. It is that approach which is discussed most fully by theorists and historians. Moreover, at present, when there is discussion about the prospect of war, it focuses on such possible conflict, for example between Russia and NATO, or China and the United States, or Iran and Israel.

However, as will emerge repeatedly (and from different directions) during the book, but requires emphasizing now, this is an insufficient approach. That is the case whether we are looking at the functional or ideological/ social/cultural criteria offered above, or if we are simply taking an empirical stance. Indeed, any definition of war in terms of a public monopoly of the use of force has to face the heavily contested nature of the public sphere, the extent to which it might not have a well-recognized and sharply distinguished nature, and the role and resilience of "private" warfare, one that is reflected in other facets such as militia forces and the extent of private fortification. Each of these points has been important historically, is significant today, and will go on being relevant. The situation is not a constant, and there are major political, cultural, and historical variations, but to focus on the public monopoly of warfare is flawed. It also does not clarify how that monopoly works.

So also with focusing on the legal codification of war, or rather, legal codifications, a process that does not cover much of the organized large-scale violence of the past. As a separate issue, the rulers of sovereign states did not necessarily declare war, which affects the question of definition and the possibilities for "hybrid" actions of some type. This has become far more the case since World War II, although the pre-invasion moves of Germany in Austria in 1938 offer an example that shows that the outbreak of that war was not a total divide. More generally, the idea of "hybrid" war as somehow new is totally wrong. For instance, England and Spain engaged in such conflict with each other for over a decade before war formally began in 1585.

A task-based understanding of warfare can be useful, but that also poses problems. Tasks are related to goals, and the latter are crucial when defining war in terms of intentionality. That, however, needs to be considered in terms of contexts, notably those in which goals are framed, communicated, and implemented. For example, the First Punic War between Carthage and Rome (264–241 BCE) apparently began as a blunder due to clashing interests in a region of Sicily and without clear, calculated goal-oriented operations on the part of either side. In eighteenth-century India, military operations were sometimes related to revenue collections and often dictated by the need to seize or protect territory. It is not easy to separate such operational aspects of war from the use of force to collect or seize revenue. The Marathas, an Indian warrior group powerful from the late seventeenth century to the early nineteenth, showed a considerable overlap in their raiding. The same point is relevant for many other societies, from Papua New Guinea to feudal Europe.

History, of course, deals with time, and its values are those of time. What that means is that, while memorialization seeks to keep alive the experience of war, as well as often partisan accounts of particular wars,[2] definitions and understandings are transient, changing, and culturally specific, all products of culture. This specificity is a matter of space and time, and that cuts across, and often invalidates, attempts to provide an unchanging definition. Indeed, a failure to note the nature and impact of chronology, and of chronological change, is labeled ahistorical. That such an attempt, however, is not seen as inappropriate by many other disciplines, for example political science, underlines the extent to which, whatever some of them may imply, historians do not own the past or its interpretation. Yet, other attempts to employ the past for models of wide applicability both utilize the work of historians and also face scrutiny in terms of the skepticism that characterizes history, and notably, and valuably so, in the Western empirical tradition.

Anthropology and collective psychology both play a role when assessing the causes of conflict. Indeed, one aspect of warfare is that of anti-societal conflict motivated by existential assumptions of threat and struggle. The treatment of enemies as beasts or as subhuman has been widespread, especially, but not only, with religious and ethnic configurations. This practice has been seen in both international and civil warfare and has served to bridge the two. In multi-generational conflicts, "injustices" were/are nurtured to justify the cause of future conflict.

This element became more prominent as nation-states replaced empires in the twentieth century, particularly with decolonialization. However, it was also the case with many empires that sought to pursue policies of ethnic supremacy and that represented ethnic goals and ideologies. The centrality of the Holocaust to Hitler's views and, finally, goals is readily apparent, and the Holocaust became a major part of German war-making in World War II. The totally one-sided war on Jews was therefore a conflict that should be considered as a war and, indeed, Hitler regarded it as a meta-historical struggle. This point is also relevant for other genocides, which generally cannot be understood outside the contexts of the strategic-level and the operational-level conduct of war; and, at once, poses problems with any idea of war as inherently legal because of the fact of sovereignty, or noble due to the test of battle. As a consequence, moreover, certain states, for example Bangladesh, do not distinguish in terms of collective memory between a genocide and a war of independence, the first being perpetrated by the target of the second, Pakistan in 1971.

The relationship between war and state development is also significant and a topic that greatly interests historians and political scientists. The topic involves the issue of what is a military institution. By many definitions, a collective entity need not have an "army" to fight a war. The cases that come to mind are traditional societies, such as hunter-gatherer groups, with little division of labor and in which all males are expected to bear arms. This point challenges the scholarship on the origins of war that ties it to state formation in the shape of an agricultural productivity high enough to permit a hierarchical social structure with significant division of labor. This account of the origin of armies was somehow tied to the development of "war."

Discussion of some of the issues is long-standing. The comparison, in book 19 of his *City of God*, by the Church-Father St. Augustine (354–430 CE), of Alexander the Great of Macedon with a company of thieves—"in the absence of justice there is no difference between Alexander's empire and a band [*societas*] of thieves"—was a moralist's vain attempt to argue that intentionality, not scale, was the crucial issue. In accordance with Christian history and teaching, and notably of the example of the persecution of the early Christians, sovereignty was not seen as a legitimator of slaughter. "Just war" was presented both as desirable and as a phenomenon that could be defined. Separately, there was a world of disorder. For example, the reference to theft brought out the key role of looting in motivation, a role that continues in some cases to the present, both at the level of individual soldiers and with regard to their states and/or commanders seeking to seize resources.

The significant point about sovereignty can be approached from a variety of directions, which can be grouped as ideological, legal, and functional, without suggesting that this categorization is precise or uncontested. If a key issue with warfare is how it is possible to persuade people to kill, and to run a strong risk of being killed, then there was not much functional difference in the sixteenth century between "state-directed" warfare and its *ghazi* (the Muslim system of perpetual raiding of the infidel) and, indeed, piratical counterparts. The organized killing of humans was central to each, even if the objectives behind this killing was different. "States" were inchoate, and not generally seen as enjoying the right to monopolize warfare and, alone, to legitimate conflict. Points similar to that of Augustine were made by later writers, both political commentators and others. The comparison of the ethics of "the state" with that of "criminal groups" can be seen in Shakespeare's *Antony and Cleopatra* with regard to pirates and in Friedrich Schiller's 1781 play *The Robbers*. The latter point was brought out in the 1921 version by the director Erich Ziegel, which used modern guns as props, included allusions to industrialization, and dressed the actors in contemporary military uniforms.

Religion was definitely (and remains) a crucial legitimator for conflict. Thus, Alfonso VI of León and Castile (r. 1065–1109), who, in 1085, captured Toledo, earlier the capital of Visigothic Spain, and the primatial see, from the Muslims, saw this as a key validator of his position. Alfonso referred to himself as "Emperor of all the Spains," and moved his capital from León to Toledo.

Today, issues of legitimacy continue to come into play, not least with notions of religion as a key legitimator, with the claim to the attributes of sovereignty by groups not recognized as such, and with the rejection of the ideas that sovereign governments have a monopoly of force. This point overlaps with the question of the distinction between military and civilian. This issue captures the tension between moves for a universal legal standard for the conduct of war and, on the other hand, as emerges repeatedly, very varied practices, some grounded in military methods and some in culturally specific attitudes. A universal legal standard presupposes that there is an agreed definition of war.

Moreover, the "trend" at present is to undermine the distinction between military and civilian. In fact, this distinction has often been slight. For example, the pursuit of the Cold War interacted, as in Central America, with local conflicts. Force was a key element in the pursuit and defense of local interests, employment, and status. It also played a major role in rivalries between, and within, communities, families, and generations. To help mold the situation, governments turned to particular groups or sides, and the articulation of the resulting tensions ensured that it was not easy to isolate or limit particular quarrels. Instead, action was taken at one level in order to influence power and struggles at another. Weapons and other forms of support were moved to allies. For example, in Mindanao in the southern Philippines, a persistent problem with a Communist insurrection led the security forces to arm private forces, such as the Kuratong Balleng gang, and they, in turn, pursued local interests, including crime and politics.

In practice, there is no one trend today, but, rather, a number of tendencies. They include the development of "hybrid warfare" and notably so by Russia. In the United States, moreover, the distinction between civilian and military functions has been clearly lessened in the case of the CIA, a civilian body with military attributes and functions. It oversees, for example, drone bombings and its secrecy makes war "invisible."[3] Targeted assassinations are an aspect of directed conflict. They are also part of "hybrid warfare." Militarized police also erode the distinction between civilians and police. In Turkey, special police and gendarmerie units, formed in 1982, have led operations against Kurdish militants since 2015, and were moved under the command of the Interior Ministry in 2016 due to governmental concerns about the loyalty of the army command. In 2018, the president took direct command of these much-expanded forces.

The use of force by (major) states against those they deem internal opponents cannot be rigidly separated from war simply because the states do not accept the legitimacy of their opponents, a point that is valid for the treatment both of genocide and of slave rebellions. Some modern commentators do not treat the latter as "wars," but, in terms of scale and intention, they were. Indeed, understanding slave rebellions as warfare throws light on the widespread assumptions of regularity in the organization and conduct of warfare, for these rebellions, instead, were characterized by the alternative practices bound up in the nature of irregular warfare and "small wars."[4]

As far as legitimacy is concerned, there was a distinction between wars begun by imperial powers, such as Ottoman Turkey, Safavid Persia, Mughal India, and Ming or, later, Manchu [Qing] China, and war within these empires, but the latter could be large-scale, more so indeed than external conflict, and could be regarded by contemporaries as war. Moreover, since each of those states rested on warfulness, war, and conquest, they had highly bellicose values. In addition, attempts to overthrow these, or other, states can be seen as war, even if they are different to conflicts between sovereign powers. So also with the far more widespread case of trying to deny the authority and thwart the power of states, although that risks making much political behavior appear as warfare.

As a related point, there is the question whether it is only the outcome that earns the designation war. For example, there is the contrast between the American and Latin American Wars of Independence and, on the other hand, the Irish rebellion/revolution of 1798 against British rule, which can be presented as a would-be war of independence. In addition, there are long-standing disagreements over whether the "Indian Mutiny" of 1857–59, as the British colonial rulers termed it, should be referred to as a mutiny, a rebellion, a war of independence, or just a war. "Pontiac's Rebellion" or "Pontiac's Conspiracy" in North America in 1763–64 is misnamed since the Delaware (tribe of Native Americans) and their allies did not acknowledge British sovereignty, while Pontiac was not even the main instigator. The politicization of wars very much accounts for their naming. More generally, this question indicates the role of "history wars" and national accounts/myths in setting the agenda for discussion and establishing definitions. The particular issue feeds directly into modern revolutionary claims about struggles as wars.

Returning to the issue of defying and enforcing authority, the absence of strong, or even any, police forces frequently ensured that troops were used to maintain domestic order and control. This is also the case in many countries today. That, again, raises the question of a definition of war as the use of force—in other words through function rather than intention.

Problems of definition continue to recur. By the spring of 2017, the conflict in Ukraine that began in 2014 had cost ten thousand lives and displaced over 1.7 million people. Russian-backed separatists proclaimed the Donetsk People's Republic in 2014 while Ukraine calls the conflict zone an area of anti-terrorist operations. The deployment of Russian tanks outside Ukraine served to put pressure on the latter, while within Ukraine, Russian forces defeated the Ukrainian army, such that a line of division was left. However, the conflict was not called a war. In May 2018, the breach by Palestinians of Israel's border fence around Gaza led Israel to use airstrikes to destroy two boats belonging to the Hamas-run marine police that were due to meet a flotilla of protest ships planning to break the Israeli maritime blockade, a blockade that Israel enforced with military action against civilian ships in both 2010 and 2016. How far this is an appropriate response and how far it amounts to acts of war are matters of controversy.

Turning to culture, the use of the concept of bellicosity in part overcomes the unhelpful distinction between rationality and irrationality in discussing motivation and conduct. Bellicosity can be regarded as crucial to both motivation and conduct, as well as both (or either) a rational or an irrational response to circumstances.[5] To refer to bellicosity as a necessary condition for and, even, definition of war, is not to confuse, as may appear the case, cause and effect, or to run together hostility and conflict, but to assert that the two are coterminous in many circumstances. Bellicosity also helps explain the continuation of wars once begun.

An emphasis on bellicosity leads to a stress on the assumptions of ruling (and other) groups, assumptions that are often inherent to their existence and role. There was certainly in the past a reading of cultural norms, including historical experience, in a fashion likely to encourage a recourse to violence. For example, in explaining success in the Americas in the sixteenth century, Spanish commentators did not choose to emphasize native support. Instead, there was a stress on cultural values, including religious support.

In contrast, there is a growing reluctance today to fight in many societies, certainly in comparison to the first half of the twentieth century. Moreover, partly thanks to growing professionalism and the abandonment of conscription in many states, the military is less integrated into society than when most men could expect that they might be called up. As a related, but separate, process, there has been a process of civilizing of the military, or rather, a degree of civilianization. It is more difficult than before in many states for militaries to act as a semi-independent adjunct of society able to follow its own set of rules. Instead, militaries are expected, in warfare, policing, discipline, and much else, to conform to current societal standards of behavior. This is a pressure that has caused scandals and court cases in the United States and Britain. Moreover, this process can be seen to lessen bellicosity.

However, as, in effect, modern versions of the English *Book of Sports*, which taught archery in the seventeenth century, the popularity of war toys, games, and films suggests that military values are still seen as valuable, indeed exemplary, by many, or at least some. Moreover, it is necessary to be cautious about extrapolating from this model of change to the entire West, or, even more, world.

In addition, as recent years have shown, what can be characterized as globalization can readily lead to the use of force by outside powers and by international organizations, notably the UN and African bodies. Linked to the question of Western interventionism, but also separate, democracies, once roused, as the United States was following the terrorist attacks in 2001, can be very tenacious in war, or, at least, their governments can be. Israel is a prime instance of this tenacity, while Britain provided an example during the Falklands War of 1982.

The suggestion that the "West" has become less bellicist might also appear ironic given its nuclear preponderance, the capacity of its weapons of mass destruction, and the role of its industries in supplying weaponry to the rest of the world. Indeed, it might almost be argued that this strength is a condition for the decline of militarism. A decline in bellicosity could also be seen as owing something to the prevalence and vitality of other forms of "aggression," for example what might be seen as economic and cultural imperialism. This approach, however, does not match the desire for some, generally observers not participants, for war to be conflict between warriors, rather than soldiers. Warriors are strong individually as opposed to soldiers, who offer organized force. Film depictions of war often focus on the former, on "one-man armies," and thus greatly misrepresent capability, and, indeed, the issues involved in organizing people to fight.

As a difference in bellicosity between cultures in the present world can be stressed, for example the Netherlands and Iran, then that suggests the continued vitality of a model of transcultural variations. This point underlines the culturally contingent nature of definitions of war, as of other phenomena. The notion of each state as having a distinctive strategic culture is in part a reflection of transcultural variations, albeit in a different fashion to those generally understood as cultural being, instead, more overtly political. Cultural factors certainly play a major role in assessments of relative power,[6] as well as in the response to risk.[7]

An emphasis on cultural contexts, including the response to risk, within which war is understood, even welcomed, as an instrument of policy, and as a means and product of social, ethnic, or political cohesion, is also, in part, a reminder of the role of choice. As such, this approach is a qualification of the apparent determinism of some systemic models. A denial of determinism also opens up the possibility of suggesting that the multiple and contested interpretations of contemporaries are valuable. This underlines the importance of integrating them into the explanatory models. As far as an emphasis on intentionality is concerned, bellicosity leads to war, not through misunderstandings that produce inaccurate calculations of interest and response, but rather from an acceptance of different interests and values, and a conviction that they can be best resolved through the use of force. As such, war can be the resort of both satisfied and unsatisfied powers: those who do not want a change in the system and those, the revisionists, who do, and especially so in terms of relative power.

This observation is also relevant for civil wars. Indeed, the latter help take us to the point of understanding the causes of war as an aspect of history as a whole rather than simply of military history. The domestic level thus emphasizes variety, as the nature of states/countries is very different, and also changes through time. Moreover, the balance between domestic and international wars varies greatly. In the case of China, its internal wars have been catastrophic, indeed human tragedies on a major scale. Their causes, however, are very diverse, including politics, ideology, regionalism, and resources in the shape of the desire to take control of what the bureaucracy could collect. More blood has been spilled on unity in China than would have been by disunity.

Across the world, despite the widespread conviction of necessity and a frequent belief in providential support, the resort to war is a choice for unpredictability, which is a matter not simply of the uncertain nature of battle, but an inherent characteristic of the very nature of war. The acceptance that risk is involved, and the willingness to confront it, are both culturally conditioned. So also is the "status dissatisfaction" that has been seen as an important element in creating pressures for changes in international relations, notably with "power transition theory."[8] An emphasis on the cultural dimension is encouraged by an analytical stress on context and agency: "the values and ideas of leaders determined the degree to which they envisaged war as an opportunity, necessity, or catastrophe," and their stance mattered.[9]

Bellicose factors in leadership and society can be related to the structure of international relations and to the perceived alternatives available to leaders in reaching their objectives in foreign relations. These alternatives owe much to the human capacity to anticipate the future in a complex fashion.[10] The structure of international relations encompasses the responses of different powers, but the extent to which the structure shapes these responses is unclear, even questionable. Indeed, this point underlines the extent to which analytical notions shift. The current habit is to focus on an international system, with the system treated as something that exists in the real world, and indeed has legal force and the capacity for political and judicial action, rather than simply as an analytical category.

In practice, the framework for analysis has changed through time. The causes of war are a topic that is classically presented as timeless, not least because Western commentators like to cite Thucydides's *History of the Peloponnesian War* as a still-valid explanation.[11] This approach, which is paralleled in the treatment of Clausewitz, the author of *On War*, has its value, because there are indeed long-term, even "timeless," characteristics in terms of both individual and collective human ambition and psychology, such as fear and ambition. At the same time, this approach risks underplaying the significance of changes through time, changes both in the phenomenon itself and, interestingly, in the explanations of it. The latter changes can be presented as matters of intellectual advance or fashion or trends, but there is a danger of downplaying the significant role of politics in analyzing the subject. Indeed, as with other topics, for example strategy and geopolitics, there can be a misleading tendency to depoliticize the analysis and, in doing so, to present a "rational" account. That process can misrepresent the causes of action and can even lend itself to a deliberate failure to acknowledge the motivations involved.

To return to the international system and to indicate the role of change, in the seventeenth and eighteenth centuries, alliances and other formal agreements between states were the focus of attention in Europe. Thus, the system was the relations between states, and these relations were the essential element. Wars were encouraged by the strength brought by alliances. In the nineteenth century, however, the system was presented in Europe as a mediating agent that constrained the activities of its member units and controlled the interactions between them and the system's environment. This concept was drawn from the model of a living body and can be termed organic or biological. Such an approach, as taken today in the West, sees the bellicosity of individual states as a problem, if not a crime, and, instead, requires a moralization of war directed against the bellicosity, a moralization not only to fulfill the goal of mobilizing public support but also to explain why conflict is necessary in some circumstances and unacceptable in others. Linked to this, past wars and their origins can be, and are, interpreted in new ways to suit the needs of the present.[12]

We can, therefore, move to the slippage in the description of warfare referred to earlier. Bellicosity is reformulated, away from an aspect of sovereignty. Instead, bellicosity becomes a problem that must be dealt with, a problem that others display. Bellicosity thus becomes a necessity for action against this problem, both in terms of preserving the international order and with reference to defending systemic norms. This is a background to modern ideas about the undesirability of warfare and, paradoxically, to support for interventionist warfare, as with Western action against Iraq and Serbia in the 1990s. Responsibility for action is transferred.

These concepts are not new, and some can be seen as the secularization of earlier religious beliefs about the undesirability of conflict between co-religionists; but their application is very much new. It rests on a concept of world government through consensual practices based on a shared sovereignty linked to defined norms that restrict the independent role of the individual state. The concept and practice of such interventionist warfare from the 1990s may well have been naïve and may prove to be short-lived. At any rate, it is unclear that these ideas help us in understanding the values of those, in past, present, and future, for whom compromise is unacceptable, force necessary, and even desirable, and war crucial to identity, mission, and self-respect.

NOTES

1. Jeremy Black, *The Age of Total War, 1860–1945* (Westport, CT: Praeger, 2006).

2. Michael Boss, ed., *Conflicted Pasts and National Identities: Narratives of War and Conflict* (Aarhus: Aarhus University Press, 2014).

3. Marilyn Young, "'I Was Thinking, as I Often Do These Days, of War': The United States in the Twenty-First Century," *Diplomatic History* 36 (2012): 15.

4. Jeremy Black, *Insurgency and Counterinsurgency: A Global History* (Lanham, MD: Rowman & Littlefield, 2016).

5. Bellicosity was not universal. Naoko Shimazu, *Japanese Society at War: Death, Memory and the Russo-Japanese War* (Cambridge: Cambridge University Press, 2009), points out that anti-war sentiment was found alongside bellicose nationalism.

6. Philip Streich and Jack Levy, "Information, Commitment, and the Russo-Japanese War of 1904–1905," *Foreign Policy Analysis* 12 (2016): 489–511.

7. Justin Conrad, *Gambling and War: Risk, Reward, and Chance in International Conflict* (Annapolis, MD: Naval Institute Press, 2017).

8. Andrew Greve and Jack Levy, "Power Transitions, Status Dissatisfaction, and War: The Sino-Japanese War of 1894–1895," *Security Studies* 27 (2018): 148–78.

9. Richard Ned Lebow, "What Can International Relations Theory Learn from the Origins of World War I?" *International Relations* 28 (2014): 404.

10. Lawrence Freedman, *The Future of War: A History* (London: Allen Lane, 2017).

11. Thucydides, *History of the Peloponnesian War*, trans. Richard Crawley and ed. Robert Strassler (New York: Free Press, 1996).

12. David Stevenson, "Learning from the Past: The Relevance of International History," *International Affairs* 90 (2014): 5–22.

Chapter Two

To 1500

ORIGINS

Violence is hardwired into the process by which many species live. It is the means of life for those carnivores, the majority, who prey on living animals, while herbivores also fight competitors and attackers. Fighting is also the way to defend life, dwellings, and food from other species, and from members of one's own. Moreover, the nature of evolutionary adaptation to circumstances and the competition this entails can involve violence. Although competition does not have to be violent, it frequently is.

It is not therefore surprising that man, an omnivore competing with other species and with other humans, resorted to violence. Fighting is not, as was argued by commentators in the 1960s who were unconsciously copying Judeo-Christian ideas of the Fall of Man due to Adam's sin, some result of the corruption of humankind by society. Instead, fighting is integral to human society. The development, both of weapons and of relevant techniques, permitted the hunting, and maybe the hunting-out, of large mammals, notably mastodons and mammoths, which were individually more powerful than humans. Thus, in North America, in 10,000–9000 BCE, early settlers armed with large stone points were able to pierce mammoth hide. Aside from killing animals, these tools made it possible to cut up their bodies and thus eat them. Hunting and fishing were more general, extending beyond large mammals. A burial pit found in Alaska in 2010 revealed the bones of an infant girl and another baby who died about 9500 BCE, laid on a bed of antler points and weapons, and covered in red ochre. In Central America, sloths, mastodons, and giant armadillos were hunted. In Europe, large mammals, such as mammoths, mastodons, saber-toothed tigers, giant deer, and the woolly rhinoceros, became extinct in the same period.

The ability to communicate through language and to organize into groups was significant in hunting these, and other, animals, and was part of a broader pattern of social development. The use of fire—probably widespread by about 80,000 years ago—was important to the human advance; it could provide protection, for example in caves, against other animals. Stone-blade technology gradually improved, culminating in microlithic flints mounted in wood or bone hafts that could be used as knives or as arrowheads. Humans were subsequently to develop metallic weaponry, culminating, with the Iron Age, in effective iron or iron-tipped ones.

Their skill in hunting had helped ensure that humans were better able than other animals to adapt to the possibilities created by the retreat of the ice sheets at the end of the last Ice Age in around 10,000 BCE, the last stage of the several advances and retreats that characterized the Ice Ages. Humans were increasingly able to confront other carnivores. They gradually drove bears, wolves, and other competing carnivores away from areas of human settlement and into mountain and forest fastnesses.

Hunting clearly played a key role in early human society. Humans fighting animals were widely depicted in early rock paintings in caves, notably in Spain. Those in Cueva de la Vieja show men with bows hunting stags; those in Altamira depict bison as well as a wild boar; those in the Cueva de la Pileta show panthers, goats, and a large fish. And so elsewhere in the world. Rock art from the Tassili n'Ajer plateau in the Sahara, dating from about 6000 BCE, shows the hunting of giraffes. In Kashmir, the Burzahama site from about 4300 BCE shows hunters and a bull. The bones of animals in early human settlements provide other evidence of this fighting, and of the human use of weapons and tools to kill and cut up animals.

The meat that humans obtained from hunting provided a high protein level that allowed the brain to grow and did not require the long processing needed to digest raw vegetables and fruit. The use of fire to cook further aided the human ability to gain energy efficiently. Despite these advances, and the spread of humanity across the globe, early humans remained few in numbers, grouped into small bands, and possessing technology whose incremental advances were measured in many thousands of years.

At the most basic of levels, humans protected themselves from the outset against the claws and teeth of predators. The human need of protection could lead to a kind of permanent rest or shelter. In this context, many of the earliest fortifications were simply natural features that provided shelter or enhanced strength. Caves, ridges, thickets of vegetation, and marshlands were all natural fortifications in which men on foot could protect themselves from more mobile opponents. Caves provided residents both shelter from the weather and protection against outflanking. In Gibraltar, the same cave systems were used by successive human species, both Neanderthals and early *Homo sapiens*. In Africa, thorn bushes long acted as palisades.

In flat regions, artificial fortifications were developed where natural features were lacking or inadequate. Natural fortifications were eventually enhanced by constructing barricades of stones and earth. Fire was used to deter animals and to provide light for fighting. Wooden palisades were created to protect and control domestic animals. In the absence of wood, stone or even earth could be used. As well as protecting against animal predators—particularly wolves, bears, tigers, and lions—fortifications had to guard against other human bands, who sought to seize livestock and the land to raise them on. Livestock proved much easier to protect by impounding it in a small shelter or fortress, than were large fields of crops. However, the development of the granary or silo, a large building in which to store the produce of the fields, created a kind of fortress for agricultural products, which both offered protection against pests and weather and provided a site to defend against raiders.

Raiders sought resources, a key source of conflict. Thus, in Tasmania, there was fighting among the Aborigines, the indigenous population, over access to resources such as ochre or other valuable materials, and clans had particular areas, which was a means to limit conflict but also a source of it as interlopers sought to gain advantage. Across the world, another "resource" that was crucial at the individual and collective levels was that of women. They were a frequent cause of conflict, and, in mythological terms, the abduction of Helen was responsible for the Greek expedition against Troy detailed in Homer's epic, *The Iliad*. In addition, capturing people for enforced labor was a goal of warfare even before the rise of empire.

Accounts of aboriginal peoples as essentially pacific, a view popular in the 1960s and 1970s when it accorded with fashionable views, in "The Age of Aquarius," of what the present and future should be, are seriously misleading. They fail to appreciate the consequences of a lack of state authority to limit conflict. For example, in northern Canada, the Dorset culture of the Paleo-Eskimos of the eastern Arctic was overwhelmed from about 1000 CE by the Thule people of northern Alaska. The Dorset people were killed or assimilated.

Fortifications of a sophisticated type followed the development in some parts of the world of states around 4000 BCE. These states clashed over the need to protect or acquire resources. This led to large-scale conflict and encouraged the walling of settlements, as in Mesopotamia (modern Iraq). The growth of agriculture, and the social and state developments that came with it, in turn allowed the accumulation of surplus resources and its allocation to large-scale fortification and the establishment of armies. Both grew ever more elaborate as the resources they were protecting became more valuable.

Whatever the nature of social development, of the political system, and of governmental practice, violence was present in several strands. It was a common feature of the human attempt to control the animal kingdom. It was also a common means of settling disputes between individuals, groups, polities, and peoples, and of dealing with outsiders. There was "public warfare," as in conflict between tribes in North America, and also "private warfare": raids with no particular sanction, often designed to prove manhood. The reliance on force reflected needs and values. Conflict appeared natural, necessary, and inevitable, part of the divine order, the scourge of divine wrath, and the counterpart of violence in the elements, as well as the correct, honorable, and right way to settle matters.

Legends and other identity accounts focused on conflict. Heroic narratives played the same role as in Homer's *Iliad*, an epic account of the Greek siege of Troy. In this account, honor was the key spur, honor in the shape of control over a woman, Helen, but also, and more consistently, of relations between men. Moreover, the gods played a direct role in the struggle, frequently in competition with each other.

ANCIENT EMPIRES

Many of the gods were warlike, with the rulers their representatives. This was true of the Assyrian empire in modern Iraq founded in 950 BCE, a formidable military force. Their gods were warlike beings, and the Assyrians worked hard to spread the domain and worship of their chief deity, Ashur. So also in the fifteenth century with the Aztec empire, which expanded from its base in the Valley of Mexico after 1427 CE. Authority was also sacred. Huitzilopochtli, the God of War, was their patron deity, and human sacrifice played an important part in Aztec culture. Like the contemporary Incas of the Andean chain in South America, the Aztecs were highly militaristic, although they also made adroit use of alliances to cement their rule. Maori mythology focused on the demigod Māui, under whom the islands of New Zealand allegedly originated. The assertion of a relationship to Māui was important as part of a process in which tribal success over other tribes involved conflict between gods. This process led to the spiritual union of conquerors with the land they had conquered.

The Assyrians' terroristic style of rule, which involved mass killings, torture, and deportation, ultimately failed because it bred hatred and fostered rebellions. This trajectory, however, was not an automatic process. The same characteristics could have been employed to describe Rome, notably in its destruction of Carthage and Corinth, but this practice succeeded in suppressing opposition. For Rome, conflict was sequential. Expansion into one area, for example northern Italy, led to new commitments, problems, opportunities, and, therefore, conflict.

War was also glorious, and created reputations, such as those of Caesar and Pompey. Both the republic and later the empire celebrated victories. Generals were granted public parades, and the gods were called forth in the cause of success. The Emperor Trajan, for example, orchestrated the dedication of his column in Rome on May 12, 113, the anniversary of the dedication of the Temple of Mars Ultor in 2 BCE to coincide with the beginning of a new war against Parthia and because the help of Mars Ultor was being invoked.[1] Similarly, Hannibal took a vow in the Temple of Baal in Carthage to take revenge against the Romans for their defeat of Carthage in the First Punic War. However, religious motives were not crucial to these wars, either the Punic Wars or other conflicts.

At the same time, Rome, like other empires, relied on cooperation, both with neighbors, including client states, and cooperation in ruling subject people, as well as on conquest, so that conflict was not continual. In addition, the conquered leadership was given the chance to become Roman, which led the elites of many cities to become loyal allies. This cooperation was economic and cultural, as well as political and military. As well as cooperation with their own subjects, empire had to work with outsiders, "barbarians" from beyond the frontier, where the careful politics of mutual advantage, and the ability to create a sense of identification, were more useful than bureaucratic organization. Chinese relations with nomadic and semi-nomadic peoples of the steppe combined military force with a variety of diplomatic procedures, including *jimi* or "loose rein," which permitted the incorporation of "barbarian" groups into the Chinese realm. Their chiefs were given Chinese administrative titles, but continued to rule over their own people in traditional fashion.

Militarization provided strength. The large size of the military of republican Rome that had enabled it to defeat Carthage and other opponents derived from the organization of the peoples of Italy into various citizen and allied statuses, all of which were required to provide military service. Like the Han rulers of China (206 BCE–220 CE), the Romans believed in a mass army based on the adult males of the farming population, which provided huge reserves of manpower. Perhaps up to a quarter of a million Italians were in the Roman army in 31 BCE, nearly a quarter of all men of military age.

A focus on expansionist warfare and on fighting for defense does not capture the full nature of imperial conflict; empires also faced serious civil wars. Those in Rome included struggles, under the republic, between politician-generals for power, seizures later of the imperial position, and slave and other revolts. The causes of these types varied, but the common element was the use of force to advance interests and settle disputes.

DISCUSSING THE CAUSES

Most of the discussion of the "causes of war" was oral and presented in terms of elemental struggles between gods, as well as with reference to the rulers on Earth to whom they were linked. The polytheistic nature of most religious systems encouraged narratives of struggles between gods. For monotheistic religions, such as Judaism, there was struggle between the "true God" and pagan cults, such as those of Baal.

Alongside the general emphasis on religious elements, there were texts that remain of interest today, notably the *History of the Peloponnesian War* by Thucydides, a fifth-century BCE Athenian general and historian. He presented the growth in Athenian strength as a destabilizing force within the Greek system. Thucydides described Athens as an economy, state, and society, transformed and empowered by maritime commercialism, and its rival, conservative and landlocked Sparta, as unable to respond: as still agrarian and reliant on slavery. This difference was seen by Thucydides, and by subsequent scholars who have used his account as the first text on the causes of war, as responsible both for a destabilizing shift in relative power and for a degree of cultural animosity that made it difficult to adjust disputes.

Modern commentators have tended to emphasize the former factors, although the latter was also very important. A common "Greekness" did not prevent important politicocultural differences within the Greek world. A comparison can be made with the nineteenth-century world, with the European transoceanic empires seen as latter-day Athenian empires. More profoundly, in Classical Greece as today, ideas of self-interest aligned with those of ethical purpose: there was a convenience of congruence that eased the path to war.[2]

Thucydides's text is currently deployed in order to argue about the likelihood or probability of conflict between a declining United States and a rising China.[3] The use of the text in this fashion raises specific issues to do with the understanding of it, including its translation, and also more general questions about the relevance of historical texts as providing supposedly timeless lessons, and about argument by analogy. Texts are employed as both analysis and prospectus, which is problematic. There are also relevant points about the extent to which the understanding of the "lessons" can itself change the applicability of the example.

ISLAM VERSUS CHRISTENDOM

A similar variety in the type of conflict can subsequently be seen with Islam. The most effective of the invasions of the Classical world were mounted by Arabs converted to the dynamic new religion of Islam. Launched by Muhammad, this movement rapidly came into conflict with the paganism that prevailed in most of Arabia, and his forces captured Mecca in 630 CE. His successors, known as Caliphs, united Arabia, defeated the Byzantines (the Eastern Roman Empire) and the Sassanians of Persia, and conquered southwest Asia and Egypt by 642, going on to advance across Persia into modern Afghanistan, as well as across North Africa, reaching Morocco in the 680s–690s. In an apparently inexorable process of expansion in all directions, Arab forces benefited from mobility and high morale. Arab soldiers were present up until the conquests of Egypt and Persia. It then became the practice for Arab leaders to use converts from the most recently conquered regions in order to move farther: Persians into Central Asia, Egyptians into Tunisia, Tunisians into Morocco, and Moors into Spain, an invasion begun in 711. They were led by Arabs, which created a problem following the battle of Tours/Poitiers in 732: the Arab leader and most of his lieutenants were killed in battle by the Franks, and the troops had little direction because they relied on that leadership for religious as well as military guidance. The Muslim advances helped to mold the modern world. It was a cultural as much as a military advance, and its progress was to be reversed in relatively few areas. Wars can be agents of creation as well as of destruction.

At the same time, expansionist warfare was not the sole type of conflict in which Islam took part. Instead, the Muslim world was itself rapidly and increasingly fractured with rebellions and civil wars. The political unity of Islam ended when a member of the Umayyad family overthrew the Abbasid governor of Spain in 756. In addition, the Fatimids, who were Shi'ites (followers of an Islamic tradition that believed political power should be held by the descendants of Muhammad's son-in-law, Ali) established a caliphate in Tunis in 910 and conquered Egypt in 969. These and other divisions led to repeated conflict. For example, in the second half of the eleventh century, the Seljuk Turks defeated the Abbasids and the Christians of Byzantium, while the Fatimids, based in Egypt, competed with the Seljuks for control over the Holy Land. In the eleventh century, the Caliphate of Córdoba disintegrated into civil war and a series of independent kingships or *ta'ifas*.

Alongside these divisions within Islam, there could be an absence of external hatred of Islam, and notably so in Western Europe, which was less threatened by attack.[4] However, certainly encouraged by the Papacy, the idea of a "war between civilizations" very much existed. Indeed, it was an obvious counterpart to that between good and evil, with good understood in terms of belief in a particular religion. At the same time, it was all too easy to find evil opponents within one's own religion. Yet, because religion was often the language of conflict, that did not mean that it necessarily caused individual conflicts.[5]

The advances by the Seljuks against the Christian Byzantines, whom they defeated at Manzikert in 1071—and concerns about access for Christian pilgrims to the holy city of Jerusalem—led Pope Urban II in 1095 to preach a holy war against Islam that became the First Crusade. It was the first of a series of Christian holy wars, inspired by the potent ideal of fighting against the external or internal foes of Christendom, and also of substantiating Papal claims to lead Christendom. The Crusaders captured Jerusalem in 1099, and they founded a number of states in the Holy Land.

Their movement led to the creation of a novel military organization that had political overtones: the Military Orders. The Templars and the Hospitallers (the Knights of St. John), warriors who had taken religious vows, had troops and castles and were entrusted with the defense of large tracts of territory and sometimes with sovereignty over it. There were other Military Orders, notably that of the Teutonic Knights that began in the Holy Land but, from the mid-thirteenth century, focused on the eastern Baltic, and the Orders of Calatrava, Alcántara, and Santiago in Spain. The rationale of these orders was that of conflict.

In Spain, the conquest of Muslim areas saw Christianization, notably with church-building and renaming. In Toledo and Seville, the cathedrals were built on the same foundations as the mosques they replaced, while that in Cordoba was built in the middle of the mosque. Victory was celebrated and used as a political tool, for example by Alfonso X of Castile (r. 1252–84). Some Muslims were expelled, others enslaved, although still others continued free to till the soil. Once conquered from Byzantium, Syria, Lebanon, Israel and Egypt saw Islamicization.

Yet, confessional warfare could include a measure of inheritance of power alongside conquest. This was seen when the Ottoman (Turkish) ruler Mehmed II conquered Constantinople in 1453, its ruler, the Emperor Constantine XI, dying in the defense of the city. As well as making their empire contiguous and gaining a major commercial and logistical base, the Ottomans won great prestige in both the Muslim and Christian worlds, and took over a potent imperial tradition. Their capital moved to Constantinople, while Mehmed took the sobriquet "the Conqueror" and the Byzantine title of "Caesar." He had grown up amid imperial ideas.

The Crusades also saw rivalry within both Islam and Christendom. For example, the most famous Muslim warrior of the High Middle Ages, Saladin (c. 1138–93), was a Kurd who rose to prominence in the service of his uncle, the governor of Egypt, whom he succeeded in 1169, and he founded the Ayyubid sultanate in 1171. Saladin's forces pushed west to Tunisia and south into Sudan and Yemen, while he also took over southern Syria, capturing Damascus in 1174. These represented age-old geopolitical goals and practices of the Egyptian rulers, both pre-Muslim and Muslim, ones that Saladin proved particularly successful in achieving. His was an adroit linkage of military, political, and diplomatic strategies. In 1187, Saladin proclaimed a jihad and crushed the Christian kingdom of Jerusalem.

Conflicts within the Islamic world reflected the range of causes of war, from ethnic tension to succession disputes, ambition to fear, a failure to handle power transitions to clashes over buffer zones. The last can be seen in conflicts between the Ottomans and their neighbors. Uzan Hasan, head of the Aqquyunlu or White Sheep confederation of Türkmen tribes who ruled Iraq, backed Ishak Bey, a contender for the throne of Karaman (south-central Anatolia), and this support led in 1464 to the expulsion of his half-brother, Pir Ahmed, the son of an Ottoman princess who had been installed there with Ottoman approval. The issue helped increase Ottoman hostility toward the Aqquyunlu, whom they defeated in 1473, having driven out Ishak Bey in 1465.

Conflict was sequential, as opportunities were created and priorities changed. In 1465–71, the Ottomans secretly backed Shah Suwār of Dhu'l-Kadr in southeast Anatolia, who had deposed his brother as emir and renounced the emirate's ties of vassalage to Mamluk-ruled Egypt. In turn, the Mamluks and Ottomans fought in 1485–91: the Ottoman conquest of Karaman in 1468–74 had removed a buffer, bringing frontier issues to the fore. Dhu'l-Kadr, another buffer, lost this status as a consequence of Ottoman annexation in 1515. This led Dhu'l-Kadr's nominal overlord, the Mamluk Sultan Qansuh al-Gawri of Egypt, to respond favorably to a Safavid (Persian) approach for an alliance. Threatened by concerted action, the Ottoman ruler Sultan Selim I the Grim marched east in 1516, unsure which of his opponents to attack. However, Qansuh's advance to Aleppo in northern Syria, under the pretext of mediating between the Ottomans and the Safavids, led Selim to turn against him, the nearest target, not least as Selim would have been unable to fight the Safavids with the Mamluks threatening his rear. Moreover, there may well have been economic motivation for war with Egypt given its position in the lucrative spice trade, as well as a desire to strengthen legitimacy by gaining control over key Muslim sites.[6]

WARS WITHIN CHRISTENDOM

Correspondingly there were tensions and conflict on the Christian side of the Crusades, and from their outset. Richard I of England was imprisoned by Leopold V, Duke of Austria and then by the Emperor Henry VI, in 1192–94 on his return from the Third Crusade. Most prominently, the Fourth Crusade (1202–4), under Venetian influence, attacked not the Muslims, but, first, the Hungarian-ruled Catholic city of Zara, a trading rival, and then, Constantinople itself. Alexius Angelus, the son of the deposed Byzantine emperor Isaac II, offered money, help with the crusade, and the union of the Orthodox Church with Rome (from which it had been divided since 1054), if his uncle was removed. This was secured by the Crusaders in 1203, but the new emperor, Alexius IV, was unable to fulfill his promises and was deposed in an anti-Western rising. This led the Crusaders in 1204 to storm the city, crown Count Baldwin of Flanders as Emperor Baldwin IX, and partition the Byzantine Empire between them. However, the new situation was far from stable. In 1261, the Greeks retook Constantinople, bringing this brief Latin empire to an end. At the same time, there was a potent ideological rivalry, that between Catholic and Orthodox Christians, a rivalry encouraged by this episode.

Venice should not be presented only in terms of war with the Turks. It also played an active role in conflict within Italy, with a series of wars with Genoa, from 1253 to 1381, as the two competed over maritime supremacy. More generally in Italy, domestic urban politics interacted with international power politics, as in Florence. Many cities fell under the control of seigniorial families who then sought to dominate their neighbors, as with the della Scala in Verona, the Este in Ferrara, and the Gonzaga in Mantua.

Individual ambition was a key driver in causing conflict. The most powerful state in northern Italy was Milan under the Visconti, of whom the most dynamic was Gian Galeazzo Visconti. In 1378, he succeeded his father, Galeazzo II, as joint ruler with his uncle Bernabò, but, in 1385, he had Bernabò seized and became sole ruler: Bernabò was killed the following year. The determined and murderous Gian Galeazzo then rapidly expanded his inheritance, in particular by seizing Verona and Padua, which made him also the most powerful ruler in northeast Italy. Visconti power was then extended into Tuscany with the acquisition of Siena. Bologna was temporarily captured, taking advantage of the weakness of Papal power.

In Iberia, rulers fought each other (as well as the Moors) as they sought to affirm and strengthen their status and power. Succession issues played a role, notably in the 1360s, with a Castilian civil war in which Aragón backed Peter I of Castile's illegitimate brother, Henry of Trastámara, who defeated and murdered him in 1369, becoming king. There was another war of the Castilian succession a century later. The claims of Isabella, the half-sister of Henry IV (r. 1454–74), were contested by her half-niece, Joanna la Beltraneja, the possibly illegitimate daughter of Henry. Isabella was backed by Aragón (she was married to the heir apparent, Ferdinand, her cousin), and Joanna by Portugal (she was married to Afonso V). The aristocracy was divided. The succession was decided by battle. Much of the conflict within Christendom related to dynastic disputes.

A TYPOLOGY OF WAR

This book adopts a typology of war that distinguishes three main types. First, there is conflict between polities that derived from different cultures, for example between Christendom and Islam. Second comes those between polities that derived from, and were waged within, the same culture. Third, and overlapping with the second, comes civil wars.

This typology hinges upon a workable concept of culture. This is not easy for a number of reasons. Concepts of culture change across time. Religion may appear to provide an obvious instance, but is not a comprehensive one. Hence a poor choice is offered: either to superimpose one particular concept upon any aspect of the past that happens to come under review, an approach that fails to note differences in concepts of culture; or, alternatively, to retain the changing concepts of culture that are to be found in the sources. The second approach reduces the communicability of what is being described. Moreover, continuing distinctions between cultures may appear in different contexts and be used for different purposes. For example, cultural distinctions can be employed for purposes, and analyses, of integration or segregation. For the purposes of this book, with its emphasis on perceptions of difference as a major cause of bellicosity, it is apparent that, however much cultures can be variably defined, there was, and is, an important and dynamic "middle ground" between them. Nevertheless, contemporaries had a strong sense of such differences, and war played a major role in responding to, as well as shaping, cultural identities.

Linked to this point comes a dominant theme, that of seeing tolerance and toleration as weakness. Not all individuals or cultures hold that stereotype, but the reverse is more the norm. That helps explain war. The willingness to appreciate other points of view ultimately rests on a belief that such an approach is not only legitimate but also necessary, indeed good. That willingness is less common in human society than the urge to fulfill assumptions and advance group interests. Moreover, the idea of the group as being the entire human species or even all of creation, the pantheist approach, is distinctly a minority one. The alternative restraint is that of a prudential acceptance of other points of view, but that approach rests on an ability to understand how to manage the situation. That also is limited by cultural factors. These points emerge from the consideration of war in the period covered by this chapter, which indeed encompasses most of human history. Furthermore, most of the key ideological systems and concepts of the present day, including all the leading religions, developed during this era.

Alongside similarities and a degree of continuity between the distant past and the present, there were important contrasts to the present situation. In particular, the paucity of state-directed regular forces owed much to the absence of powerful sovereign authority across much of the world. Instead, it is commonly more appropriate to think of tribal and feudal organization, rather than a state-centric system. This, and the resulting diversity in the political background to military activity, was evidence of the vitality of different traditions, rather than an anachronistic and doomed resistance to the diffusion of a progressive model. In other words, "best practice" varied, and was largely set by natural and human environments, rather than being some unitary concept dictated by weaponry and doctrine.

FIFTEENTH-CENTURY CHANGES

It is inappropriate to treat the medieval centuries as if they were an inconsequential precursor to a modern history that was supposedly begun in the late fifteenth century, maybe by the rise of "gunpowder empires," such as those of the Ottoman Turks and Mughals, or by European voyages of exploration, or, as far as the Western system was concerned, by the Italian conflicts that started in 1494. Instead, there was, as in most other respects, a continuity in the fifteenth century in the causes of conflict, rather than a change. China's most important struggle was with Mongol attackers, and Christendom's with the Ottoman advance. In the New World, the Aztecs and Incas greatly expanded their power. None of these saw new patterns of activity or causation.

Moreover, given the subsequent attention devoted to the Italian Wars as the dividing point between Medieval and Early Modern and therefore as a start to modernity (prefiguring the Peace of Westphalia in 1648), it is worth noting the role of established patterns of behavior in their causation and course. The Turkish challenge to Italy after the fall of Constantinople in 1453 led Pope Nicholas V (r. 1447–55) to try to bring peace in Italy and then form an anti-Turkish league. By the Peace of Lodi (1454), Italy's leading powers—Milan, Venice, Florence, Naples, and the Pope—recognized each other's boundaries and laid the basis for more than two decades of peace. The War of the Milanese Succession was ended, the recent major expansion by Florence and Venice consolidated, and the attempt by René of Anjou, with the support of France and Francesco Sforza, to revive the Angevin claim to Naples was thwarted.

Italy experienced a quarter-century of relative calm until Venice went to war with Ercole I, Duke of Ferrara from 1480. Although it faced a coalition of almost all the other Italian states, the settlement that ended the war in 1484 left Venice with some minor gains. More ominously, the Venetians had tried to induce Charles VIII of France to invade Italy with the promise of their help to conquer Naples.

This attempt foreshadowed the French invasion of 1494, an invasion that launched what were to be called the Italian Wars. These wars reflected not only the divisions of Italy, but also a new, or rather renewed, willingness of outside rulers to intervene, the renewal again indicating continuity with the "Middle Ages," notably of intervention by Carolingian, Salian, Hohenstaufen, Angevin, and Aragonese rulers. Initially the most important was Charles VIII, who benefited from France's earlier success against England in

1449–53 and the overthrow of Charles the Bold of Burgundy in 1477. Thus, as so often, warfare in an area, in this case Italy, was in part a product of the fate of great-power rivalries, both elsewhere and in Italy. The Balkans from the late nineteenth century was very similar.

Italy provides a good case study of unexpected conflict leading to unpredictable outcomes. The balance of forces within Italy was disturbed, and external powers returned to the peninsula, when Ludovico Sforza, Duke of Milan from 1494, asked Charles VIII for help because he feared that Alfonso II of Naples (r. 1494–95), brother of Ferdinand of Aragon, was fomenting an Italian alliance against him. Charles responded promptly, arriving in Italy in the early days of September 1494 with an army composed of 30,000 troops supported by 150 cannon. Charles captured Naples in 1495 at the close of a campaign that had led to the expulsion of Piero II Medici from Florence in 1494. Charles asserted the Angevin claim to Naples and also claimed that he was providing a base for a crusade to the Holy Land.

Some Italians welcomed his arrival. However, his success aroused opposition both within Italy, where there was growing suspicion that he aimed to seize the entire peninsula, and from two powerful rulers who had their own ambitions to pursue: Maximilian I, the Holy Roman Emperor, who ruled Austria and the other Habsburg territories, and Ferdinand of Aragon, ruler of Aragon, Sicily, and Sardinia, half-brother of Ferrante (also called Ferdinand I) of Naples (r. 1458–94), and husband of Isabella of Castile. Ferdinand, who combined Italian interests with Spanish resources, joined Venice, Ludovico Sforza, and Pope Alexander VI (r. 1492–1503) in the League of St. Mark against Charles, an alliance that reflected the rapid changes in alignment that characterized power politics. Spanish troops moved from Sicily into mainland southern Italy in 1495, and the French were driven from Italy in 1495–96. This success further encouraged Ferdinand's interest in southern Italy.

The French were not prepared to see defeat in 1495–96 as final. To do so would be to surrender royal *gloire* and to compromise regal status. Charles's successor, Louis XII (r. 1498–1515), invaded the Duchy of Milan in 1499, claiming the duchy on the grounds that his grandmother had been a Visconti, the former ruling house. The Venetians also invaded the duchy. The major positions fell to the French in August–September, while Genoa accepted a French governor. French troops were also sent to help Alexander VI's illegitimate son, Cesare Borgia (c. 1476–1507), in the area of the Romagna where he sought both to consolidate the Papal position and, in the end unsuccessfully, to create a new principality for himself.

Such ambitions reflected the degree to which territorial units were not fixed, a situation that encouraged the quest to establish new ones. That was a situation seen more generally, and one that throws light on the attempts in the modern, postcolonial, world, to retain colonial territorial borders, although, as South Sudan shows today, that attempt has not always been a successful strategy. So also with the aftermath of the fall of the Soviet empire and Yugoslavia in the early 1990s, notably with Crimea and Kosovo.

Disaffection with French rule led to a rallying of support to Ludovico Sforza, who regained Milan in February 1500. In turn, the strength of the French response and the evaporation of Swiss support led to the collapse of Sforza's army. Swiss mercenaries, some trained and effective halberdiers, the heavy infantry of the age, others pikemen used to protect handgunners, were key military players, but they had to be paid, an aspect of the range of "players" in conflict. Ludovico was sent a prisoner to France, where he died in 1508, and Louis's power was reimposed in Milan. Louis and Ferdinand partitioned Naples by the Treaty of Granada (1500), and the French captured the city of Naples in 1501.

However, as so often, the new territorial order proved precarious, thus providing an opportunity for renewed conflict, conflict that can be counted as new or as renewed depending on the assessment made. This underlines the difficulties with statistical analyses of conflict. Disputes with Ferdinand in 1502 led the French to try to take the entire kingdom, only to be heavily defeated by the Spaniards in 1503. As a result, in 1504, Naples entered a personal union with the kingdom of Aragon under Ferdinand, while Louis renounced his claims to Naples by the Treaty of Blois of 1505.

Although Maximilian I's German troops played a key role fighting the French in the early stages, France and/or Spain increasingly dominated Italy. These were the only powers with the resources to support and sustain a major military effort. Pope Julius II, a *perpetuum mobile* of Papal plotting, formed the League of Cambrai in 1508, with Milan, Austria, Mantua, Savoy-Piedmont, the Swiss, and France, in order to attack and despoil Venice. France's role was decisive, notably in defeating a much smaller Venetian army at Agnadello (1509). Indicating the consequences of defeat in battle, much of the Venetian *terraferma* then rebelled, only for Venice, in this crisis, to display resilience and regain control.

Italian rulers adapted, if possible, to foreign invaders and sought to employ them to serve their own ends. For example, the Este family, the rulers of Ferrara, aligned with the French against Papal expansion. Thus, there was no inherent conflict between these rulers and foreign powers. The latter were able to find local allies, but these, in turn, could affect the relationship between France and Spain, a pattern later seen in the Cold War between the United States and the Soviet Union. In 1511, there was one of the dizzying changes of alignment typical of the period, and notably so in Italy, changes that Italian rulers helped to cause but from which they then suffered. In this case, Julius II formed the Holy League, with Spain, Venice, and Henry VIII of England, in order to drive the French from Italy. The French beat the Spaniards at Ravenna (1512), but Swiss intervention against France, and opposition to the French in Genoa and Milan, helped the Spaniards to regain the initiative. The French retreated across the Alps, while the Spaniards overran Tuscany, and Massimiliano Sforza was installed in Milan. Local politics was defined by big-power confrontations. In 1513, the French invaded across the Alps anew, supported by the Venetians, who feared Milanese expansion. The Swiss came in to help Sforza against the French, but were defeated at Novara, and the *terraferma* was ravaged by the French to the shores of the Venetian lagoon. The trajectory of international relations and conflict in 1938–41 was to show similar dizzying changes, albeit even more so due to the role of ideology.

Soon after coming to the French throne, the dynamic Francis I (r. 1515–47) invaded the Milanese anew, defeating the Swiss at Marignano (1515). The French then occupied Milan until 1521. However, the election in 1519 as the Holy Roman Emperor Charles V, of Charles I of Spain, heir of Ferdinand, Isabella, and Maximilian and of the Aragonese, Castilian, Habsburg, and Burgundian inheritances, seemed to confirm the worst French fears of Habsburg hegemony: Spain, Naples, Sicily, Sardinia, the Low Countries (Belgium and the Netherlands), and Austria were now under one ruler. Francis declared war on Charles in 1521, only to lose Milan (1521) and be defeated at Bicocca (1522). This led Venice to turn to Charles (1523), and, in early 1524, it helped his forces drive out a fresh French invasion of Milan.

A renewed French invasion later in the year led to the French capture of Milan and to alliance with Venice and Pope Clement VII (r. 1523–34). The Spaniards, however, crushed the French at Pavia (1525), capturing Francis, who agreed to peace on Charles's terms, enabling Charles to invest his ally Francesco Sforza with the Duchy of Milan. Once released, Francis repudiated the terms, claimed, with reason, that his agreement had been extorted, and agreed with Clement VII, Francesco Sforza, Venice, and Florence to establish the League of Cognac (1526), which led to the resumption of war. Again, rapid changes in alignment were the order of the day. The French invaded Italy anew, but repeated failures led Francis to accept the Treaty of Cambrai (1529) and abandon his Italian pretensions. It is unclear why these conflicts, the Italian Wars, should be regarded as more "modern" in their causation than those of many centuries before.

CONCLUSIONS

The emphasis on dynasticism, royal claims, and competitive prestige linked the period of the Italian Wars with what came earlier (and later), and Christendom with the remainder of the world. For example, when Timur the Lame (1336–1405), *amir* (general) of the Chagatai khanate from 1370, who sought to restore the Mongol empire of Chinggis Khan, attacked the Golden Horde in the late fourteenth century, prestige and legitimacy were important. Timur wished to demonstrate his superiority over Toqtamysh, the Golden Horde khan, who had earlier gained his position as head of the White Horde in 1378 as a protégé of Timur. Timur's victory in 1395 was followed by the sacking of the capital, Sarai Berke, and Timur, who preferred a systematic attempt to gain resources, and not disorganized plunder, rerouted the central Asian trade routes to converge on his capital, Samarqand. As so often, a number of factors played a role, and it is all too easy to place an emphasis with reference to present intellectual preferences.

Alongside specific factors, the scope of power and the succession of authority were long-term issues in dispute, and across the world. For example, in fifteenth-century Japan the leading *shugo* (military governor) families were hit by succession disputes that reached the stage of open warfare. The Ōnin War of 1467–77 arose from a succession dispute within the Ashikaga (shogunal) house, which was exacerbated by the role of rival *shugo*. This conflict led into the local and regional conflict that characterized much of the age of *senmgoku* (the Country at War) from 1467 to 1568.

Legitimacy was not necessarily the key issue. In some contexts, conflict itself was the central means of the polity: of its ideology, purpose, and economy. For example, in the Niger Valley in West Africa, Sonni Ali (r. 1464–92), ruler of Songhai, fought every year of his reign. Some of his campaigns, most obviously against the Tuaregs of the Sahara, who used raiding to benefit from the wealth of settled societies, can be characterized as defensive; but war was central to the Songhai state. Expeditions expanded power, maintained control over tributaries, and yielded resources, including slaves. So also earlier. For example, in the thirteenth century, the Banu Marin advanced from the fringes of the Sahara to overrun Morocco. There was scant sense of change in the causes of war in the fifteenth century.

More generally, there was only a limited, if any, distinction between war and peace. Throughout the Middle Ages, violent organized conflict—on a scale commensurate with an essentially local society—was a constant, with the result that the making of a peace did not always end it. And for long periods, war was regarded as part of the dialogue between lord (whether or not a king) and vassals, along with other modes of behavior. In the early fifteenth century, with English armies at the gates, Parisian citizens complained that aristocratic feuding was weakening the kingdom, only for the Duke of Berry to respond: "We fight each other when we please and we make peace when we see fit."[7]

NOTES

1. M. Beckmann, "Trajan's Column and Mars Ultor," *Journal of Roman Studies* 106 (2016): 124–46.

2. Peter Hunt, *War, Peace, and Alliance in Demosthenes' Athens* (Cambridge: Cambridge University Press, 2010).

3. Graham Allison, *Destined for War: Can America and China Escape Thucydides's Trap?* (Boston: Houghton Mifflin, 2017).

4. Nic Morton, *Encountering Islam on the First Crusade* (Cambridge: Cambridge University Press, 2016).

5. Brian Catlos, *Kingdoms of Faith: A New History of Islamic Spain* (London: Hurst, 2018).

6. M. Mazzaoui, "Global Policies of Sultan Selim, 1512–20," in *Essays on Islamic Civilization*, ed. D. P. Little (Leiden: Brill, 1976), 224–43.

7. Quote kindly provided by John France.

Chapter Three

War, 1500–1650

CONFLICT BETWEEN CIVILIZATIONS, 1500–1650

Wars between cultures did not arise as a simple consequence of difference, but it was far harder to solve difficulties in such cases. In explaining conflict in this (and other) periods, it is possible to focus on difference and particularly with the long-standing contrast between settled agrarian societies and their nomadic pastoral counterparts. Fear of the horseman, of the raiders from the steppe, played a major role in the consciousness of Eurasian settled societies. This fear looked back to over a millennium of repeated attacks and was, in part, a testimony to the military potency of, in particular, mounted archers. The horse-people acquired muskets, but continued to attack. For example, in 1526 the Mughals from Afghanistan under Babur brought down the Lodi Sultanate of Delhi, while, in 1722, the Afghans overthrew Safavid Persia and, from the 1750s, successfully and repeatedly invaded northern India. Many Eurasian wars thus seem an inevitable product of clashing sociopolitical systems.

This approach has been challenged by work that emphasizes interaction, with conflict only one part of the interaction. Furthermore, nomadic attacks frequently arose because the commercial and other relationships had been disturbed or the terms were no longer acceptable to one party. In short, it has been argued, the attacks were not the "natural" characteristic of the relationship, but a product of its failure, as with the Ming refusal to trade with the Mongols in the mid-sixteenth century. This refusal has been traced to Ming xenophobia and a determination to appear strong, the last a frequent product of difference and cause of conflict. In China, some officials in the sixteenth and early seventeenth centuries opposed any conciliation of Mongols and Manchu as a capitulation to "barbarians."

Rethinking the relationship implies changing the understanding of the "barbarians." No longer are they seen as product, and part, of the inchoate "other," against which civilization must defend itself, but, instead, as part of the world of civilization. If a rationality, other than that of the most basic, is ascribed to "barbarians," this can be linked to a reevaluation in which they enter into wars, rather than being in a permanent state of war.

Similar approaches can be taken in the case of cross-cultural conflict. Key instances include those between Christendom and Islam, or between the (Muslim) Mughal empire and the Hindu polities of southern India. Linked to this, it is necessary to complement an understanding of long-term antagonisms with a realization of the degree to which these antagonisms did not preclude periods in which conflict was localized and largely limited to frontier differences and groups.

That was not, however, the ideological message of either side. In Islam, the notion of jihad, war with the infidel, was actively propagated while, correspondingly, the shedding of Muslim blood was prohibited—not that this prevented conflict. In 1521, Süleyman the Magnificent (r. 1520–66) responded to the Ottoman army's preference for war with the Hungarian "infidels" and the loot they offered. By capturing Belgrade (1521) and Rhodes (1522), he succeeded where his great-grandfather, Mehmed II, the conqueror of Constantinople, had failed, and thus established his position. Comparative success within the dynasty was important to fame, and such fame was a lubricant of obedience. The issue of "outperforming" a predecessor was at once psychological and political. It also ranged across history: Alexander the Great of Macedon sought to outperform his father, Philip, Henry V of England to match Edward III, and the Emperor Wilhelm II of Germany to match his grandfather.

In the Horn of Africa, Imam Ahmäd ibn Ibrahim al-Ghazi conquered Adal (centered on modern Djibouti) in the mid-1520s after a struggle with the Sultan of Harar, a fellow Muslim, but then launched a holy war against Christian Ethiopia, which had long expanded into Muslim lands, forcing conversion on local people. Thus, as more generally, conflict continued the pattern and dynamic of earlier rivalries.[1] He sought to replace the Christian empire of Ethiopia.[2] At the same time, more than holy war was involved: Ahmäd's attempt to stop the Muslim towns paying tribute to Ethiopia was part of an economic motivation that included a search for loot and slaves. Enmity with local Muslims as well as Christian powers was also seen in northwest Africa.

In Christendom, there was a continual sense of an Islamic threat, and Crusading language remained significant into the nineteenth century, as seen with the French conquest of Algeria from 1830. The perception of menace led not only to military and diplomatic measures, for example Spanish expansionism in North Africa, but also to domestic counterparts, most obviously the Spanish expulsion, first, of Muslims and later, in 1609, of the Moriscoes, converts to Christendom who were regarded as a potential fifth column. The seizure and conversion of Muslim mosques and graveyards was part of the process. It also testified to the role of sacred landscapes in conflict. Yet, there was not always such an approach. The Ottoman Turks took a very different stance toward their Christian and Jewish subjects.

Any sense of continuous cross-cultural conflict must be tempered by the multiple commitments of rulers and their consequent need to choose between different policy options and to revisit their choices often. Despite their ideal of world empire, and the belief that a permanent state of war existed between Islam and the infidel, there was nothing inevitable for the Ottomans about conflict with Austria, Spain, or Venice. Instead, until 1639, Safavid Persia was frequently a more pressing opponent, prefiguring bitter modern rivalry between the Sunni (the majority Islamic denomination) and Shi'a worlds. There were periods of war but others of peace with Christian powers, and diplomatic relations existed, notably with Venice.

Also within the Islamic world, Babur, the Mughal leader, focused until 1514 on conflict with the Uzbeks (who were Muslims) in an attempt to maintain areas of family power, and he only turned to the Punjab thereafter. Success there helped compensate psychologically and in prestige terms for the loss of his homeland.

Ottoman Turkey was an expanding power, keen to obtain territory in order to provide fiefs, and with an ideology that glorified war. This linked the elite in a wide-ranging empire that represented a very different polity, and military, from raiders who essentially sought booty, especially slaves and other loot that could be easily moved. Some of these raiders were closely linked to the Ottomans, notably the Crimean Tatars and the Barbary corsairs of North Africa. So also with European expansion overseas: the formal goals of Portugal, Spain, France, England, and the Dutch were different to those of adventurers, although there was also a significant overlap, which in part reflected the need to rely on subcontracting and individual initiatives in order to provide, and provide for, military force.

WARS WITHIN CULTURES

The distinction between transcultural and intracultural conflicts is somewhat arbitrary, because so much depended on the perspective of participants. To Europeans, wars between Islamic powers were intracultural. This was indeed the case with the Ottoman-Mamluk wars, such as those of 1485–91 and 1516–17, which were essentially struggles for hegemony in the Near East, or the struggle between the Mughals and the Lodi Sultanate of Delhi in the 1520s for control of northern India. Issues of legitimacy and prestige, as well as stability, were posed by relations between Muslim rulers. These issues brought together dynastic, ethnic, political, and religious concerns. Each took a part in the tension between Ottomans and Mamluks. Notions of legitimacy played a role for Babur; he saw northern India as his rightful heritage as a descendant of Timur, who had conquered it at the beginning of the fifteenth century.

However, as an instance of a conflict within Islam between different cultures, the Sunni and Shi'a divide between Turkey and Persia was important, and continued, including in the war between postcolonial Iraq and Iran in 1980–88. The Safavids, a militant Shi'a Muslim religious order that conquered Iran (Persia) and Iraq in 1500–8, were led by Isma'il, who was seen as having divine attributes and deserving absolute obedience. He was regarded as the reincarnation of Imam Ali, Muhammad's son-in-law and the founder of Shi'a Islam, or as the hidden Imam, a millenarian figure. Conversion to Shi'ism helped provide coherence to the tribes supporting Isma'il. The conquest of Iraq brought the prestige of gaining major Shi'a shrines, notably Karbala, while Sunni shrines were desecrated and prominent Sunnis were slaughtered. When Herat was captured in 1511, attempts were made to introduce Shi'a rites. In addition to the general challenge to the Sunnism of the Ottoman sultans, support among the peoples of eastern Anatolia for the Safavids and for their millenarianism threatened Ottoman control and security and their sense of religious identity, while in 1505 Isma'il claimed Trebizond and thus a coastline on the Black Sea. Attacking Persia in 1514, Selim I obtained a *fatwa* declaring his opponents heretics.

Similar points about intracultural warfare could be made about conflicts between Protestant and Catholic powers from the sixteenth century, and between both and Russian Orthodoxy, notably between Catholic Poland and Muscovy. All typologies, indeed, entail a measure of arbitrary differentiation. As far as wars within, as opposed to between, cultures are concerned, the emphasis is on shared conventions and parameters of behavior in diplomacy and war. Thus, in India, the process of aggregation and disaggregation by which alliances and armies expanded and collapsed was far easier in areas of similar culture, such as the Deccan or Rajputana, than across cultural divides, for example between the Mugals and the Ahom of the Brahmaputra valley.

The Manchu conquest of China in the 1640s and 1650s, however, underlines the difficulty of assessing such issues. By then, the Manchu, who had been fighting the Ming from the 1610s, had incorporated sufficient Chinese governmental practices, and allied with enough senior figures in the Ming system, to qualify any clear-cut sense of distinction and contrast. This process was encouraged by the overthrow of the Ming emperor in 1644 by a large-scale rebellion within China.[3] However, at present in China, the Manchu legacy is rejected by some commentators seeking to develop a Han Chinese consciousness.

In China and, to a lesser extent, India, there was a major power that dominated regional political and ideological culture. It was still necessary to adjust to other powers, as the Chinese had to do, including, variously, with Japan, Korea, and the polities of Indochina. However, geopolitical dominance, and a sense that invulnerability was normal, were linked to assumptions about the proper operation of international relations.

The nature of international relations and warfare was different in Christian Europe where, despite the efforts of the Habsburgs in the sixteenth and early seventeenth centuries, there was no hegemonic dynasty or power, and where, instead, the theory and practice of a balance of power developed. Indeed, the multipolarity of the European states system, the highly competitive nature of European power politics, and the kaleidoscopic character of alliances there all help to account for the development, then and later, of "realist" accounts and paradigms of international behavior. European and European-American experiences and conceptions of international relations were to dominate subsequent writing on the subject.

The absence of such a hegemonic power in Europe was entrenched as a result of the Protestant Reformation in the sixteenth century as well as of the blocking of the Turkish advance. Thanks to the Reformation, and the subsequent, but only partly successful, Catholic Counter-Reformation, the early-modern period in Europe can in part be defined as the age of religious wars, although France, Spain, and Austria jostled for the position as principal Catholic power.

Moreover, the expansion of the Turks and the conflicts between Abyssinia (Ethiopia) and Adal in the Horn of Africa, and between Portugal and Islamic powers, such as Egypt, Morocco, Aceh in Sumatra, and the Turks, emphasize the degree to which religious conflict was widespread. This was also the case with areas not generally seen in terms of religious conflict. Takeda Nobunaga, the unifier of Japan, employed particular brutality in the 1570s to extirpate the forces of religious groups, such as the True Pure Land Confederacy, seeing their opposition as more serious than that of other warlords. The confederacy's followers formed armies known as *ikkō-ikki*, who fought to assure their rebirth. The subsequent drive by the Tokugawa Shogunate to extirpate Christianity led to an unsuccessful Christian rebellion in Kyūshū in 1637–38. Sects were important in China, notably with the 1587 rebellion in Shandong backed by the White Lotus-Maitreya sect.

Religious differences alone were not responsible for the high level of conflict in sixteenth-century Christian Europe, but they contributed greatly to it, and religion was regarded by contemporaries as a decisive factor in causing struggles between and within states, and in both long-lasting and smaller-scale conflicts. Wars were rarely fought for exclusively religious reasons, but religion was important in transforming local conflicts into either general European ones or, at least, into national conflicts. In particular, religious differences focused tensions within states by providing an element that could not be compromised. This became the killing ground.

This process could be readily seen in the German lands. In the late fifteenth century, politics and warfare in northern Germany were only loosely connected to events in southern Germany. With growing confessional antagonism, at least from the 1530s, every local or regional conflict in which opponents who subscribed to different confessional options were involved assumed a more widespread importance. Moreover, such a situation was frequently the case with local conflict. Alliances were formed comprising many, if not all, members of a confessional group nationwide. In Germany, Scotland, France, and Switzerland, confessional antagonism helped create a national framework for conflict, as opposed to a regional one. As a consequence, outcomes were far from decisive or predictable. In addition, conflicts, and their consequences, in one particular area, could affect not only the situation more widely, but also the prospect of their combination.

Had the Protestant north-German princes been conclusively defeated by the Habsburg Emperor Charles V, as indeed appeared possible in 1547, then the Reformation might have been seen in a pattern of medieval politicoreligious disputes, for example as a more serious version of the early-fifteenth-century Hussite Rebellion in the modern Czech Republic, and Western Europe might well have had an effective hegemonic power. Bullion (gold and silver) from Spain's New World conquests could have been used to sustain the traditional universalist aspirations of the Holy Roman Emperors, which would have been a functional hegemony created by the combination of a long-standing ideology and new economic, or at least fiscal, resources.

However, the burden of pressures and commitments proved too great for Charles's *imperium*, in part because of serious divisions among the Catholic powers, just as his earlier success had reflected divisions among the Protestants. The defeat of the German Protestants at Mühlberg in 1547 was reversed five years later. French willingness to support the German Protestants militarily in the early 1550s, and also to ally with the Turks, set a pattern for the early-modern period in which warring blocs did not equate with religious divisions.

Most significantly, in the Thirty Years' War (1618–48) religious differences helped to cause and exacerbate disputes, but they also complicated and cut across them: the French backed the Protestant Dutch against the Spanish Habsburgs, as well as the Protestant Swedes against the Austrian Habsburgs, before fighting Spain and Austria itself. In 1627, the end of the direct male line of the Gonzaga family produced a contested succession for the Duchies of Mantua and Monferrato. Spain, yet again concerned about the wider strategic situation of particular Italian territories, intervened in force in 1628 in what became the War of the Mantuan Succession, and France, in turn, sent in an army on the other side. So also in 1629, in support of Spain, did Austria.

As a reminder that politics in one particular area were an aspect of wider strategies, these interventions were determined by other events, for example French success against the Huguenots and that of Austria against the Danes. The Swedish invasion of northern Germany in 1630 led to the negotiation of peace in northern Italy in 1631, only for war to resume there in 1635 as part of a large-scale conflict between France and Spain. There is therefore the contrast between explaining war in 1635 and accounting for a wider pattern of hostilities that for France began formally in 1628.

The 1650s also saw crosscurrents. The "Rump," the Protestant republican parliamentary regime, already bellicose and militaristic in the aftermath of the British civil wars, attacked the Protestant republican Dutch in the First Anglo-Dutch War of 1652–54. Later in the decade, the Rump's successor, the Protectorate under Oliver Cromwell, allied with France against Spain, another Catholic, monarchical, and legitimist regime. At the same time, the conflicts of the 1650s illustrate the complexities of causation. The First Anglo-Dutch War has traditionally been explained as a mercantilist struggle largely due to commercial and colonial rivalry. However, it is also possible to consider the war in terms of the hostility of one Protestant, republican regime toward a less-rigorous counterpart, with ideological factors thereby to the fore.

That ideological divisions existed within cultural areas, but did not prevent alliances across them, ensured that there is a degree of similarity between transcultural and intracultural wars. This puts an emphasis (but not necessarily *the* emphasis) on shared functional characteristics. In these terms, the bellicist nature of societies was important in the cause of wars, as was the accentuated role of prominent individuals that was the consequence of dynastic monarchy. An automatic habit of viewing international relations in terms of glory and honor was a natural consequence of the dynastic commitments and personal direction that a monarchical society produced. This habit reflected traditional notions of kingship and was the most plausible and acceptable way to discuss foreign policy in a courtly context. Such notions also matched the heroic conceptions of royal, princely, and aristocratic conduct in wartime. This was true across cultures, and can be seen for example in Christendom, Islam, and Japan. Past warrior-leaders were held up as models.

Similarly, aristocrats looked back to heroic members of their families who had won and defended nobility, and thus social status, indeed social existence, through glorious, honorable, and honored acts of valor. Renown was important and was celebrated in display and the arts. These traditions were sustained, both by service in war and by a personal culture of violence in the form of duels, feuds, and displays of courage—the same sociocultural imperative underlying both the international and the domestic spheres, between which there was no separation in this respect. This imperative was far more powerful than the cultural resonances of the quest for peace. The peace-giver was generally seen as a successful warrior who overcame rivals, and not as a royal, aristocratic, or clerical diplomat. The Reformation and the Wars of Religion, which underlined the importance of conflict, may have strengthened this emphasis.

The pursuit of land and heiresses linked monarchs to subjects. As wealth was primarily held in land, and transmitted within the family by inheritance, it was natural at all levels of society for conflict to center on succession disputes. Peasants resorted to litigation, a method that was lengthy and expensive, but to which the alternative was largely closed by state disapproval of private violence. Rulers resorted to negotiation, but there was no adjudicating body. This led to frequent struggles, some of which were called wars of succession. For example, in 1642–44, Tuscany, Modena, Venice, and Parma joined against Pope Urban VIII (r. 1623–44) in the Castro War over the fate of Italian fiefs and the extent of Papal pretensions. The war was over possession of the Duchy of Castro, a Farnese territory but claimed by the Barberini Papacy in lieu of unpaid Farnese debts. The duchy was envisaged as a potential Barberini duchy, giving the family the same dynastic status as had been obtained by the Farnese family in the 1540s thanks to the acquisition of Parma-Piacenza. Urban VIII thus continued the practice of pursuing family benefit from holding the Papacy. In the event, the Papal forces were heavily defeated.

The need for a speedy solution once a succession fell vacant encouraged a decision to fight. That functional factor was linked to a normative one in the shape of most of the royal and aristocratic dynasties ruling and wielding power owing their position to the willingness of past members of the family to fight to secure their succession claims.

Such warfare was not restricted to the initial establishment of dynasties. In addition, individual rulers could have to fight to gain or retain their thrones, as with the Ottoman ruler Selim I in 1512, a process that caused conflict. The seriousness of a rebellion by Shah Kulu, a Safavid proselytizer, discredited Sultan Bayezid II and provided an opportunity for Selim to rebel and overthrow his father. Selim killed his brothers, which added to tension with the Safavids of Persia, who had backed one of them against him. Similarly, in India in 1658–59, savage warfare between the four sons of the Mughal Shah Jahan (r. 1624–58) led to the victory of Aurangzeb. This warfare was important to subsequent conflict with "foreign" powers; Aurangzeb was thereafter able to focus on external threats. As such, it parallels the end of the *Frondes* in France in 1652 and of the Rebellion of the Three Feudatories in China in 1681, although there could also be a continuum, rather than a sharp divide, between external and internal (foreign and domestic) challenges. Thus, the *Frondeurs* had been backed by Spain, then at war with France, while the French supported the Catalan rebels against Spain.

Although the peaceful succession of new dynasties, eased by the absence of a direct heir, could take place, as in England in 1603, conflict in such cases was generally still feared. Indeed, war and inheritance were often two sides of the same coin. This was a problem exacerbated by varying and disputed succession wars. Furthermore, the interdynastic marriages employed to give more stability to peace treaties, such as that of the Pyrenees (1659) between France and Spain, could also give rise to disputes over inheritance and thereby lay the basis for future war, with France invading the Spanish Netherlands from 1665, leading to the War of Devolution. Eventually, Louis XIV's marriage to a Spanish princess gave rise to the War of the Spanish Succession (1701–14). The bellicist character of court culture and the fusion of dynasticism and *gloire* encouraged a resort to violence in the pursuit of such interests and claims.

On a long-established pattern, warfare created polities, and the rivalries between these polities were inherent not only to their boundaries but also to their very existence. Examples include the importance of the *Reconquista* to Portugal and Spain, and to the Dutch Republic of its struggle for independence from Spain. "State"-building generally required, and led to, war, and also was based on medieval structures and practices that included a eulogization of violence. War was very important, not only in determining which dynasties controlled which lands, or where boundaries should be drawn, but in creating the sense of "us" and "them" which was, and is, so important to the growth of identity and of any kind of patriotism. War forwarded royal war propaganda, taxation, and xenophobia. The prehistory of the modern state, and its subsequent trajectory, offered much to the identification linked to war and confrontation. Indeed, a sense of identity was more crucial than specific political or constitutional structures. The construction and territorialization of identity were not restricted to the political elite. Moreover, in raising forces and persuading people to fight, the development of identity in this fashion served as an alternative framework and ideology to those summarized by the term feudal.

Rulers, however, remained in charge of the process. Indeed, the martial configurations of new and developing political communities largely remained under monarchical control and direction, a pattern that has continued with the monarchical aspects of presidential systems, from Oliver Cromwell in England in 1653–58 to Vladimir Putin in Russia today. As a related point that underlines the importance of monarchical *gloire* in the cause of conflict, war and military service were part of a code and practice that served to integrate heterogeneous groups into the gradually emergent "states" in a way that became visible in the title page to Thomas Hobbes's book *Leviathan* (1651), with the image of a mighty monarch composed of humans.

Wars reflected not only the concerns of rulers but also the ambience and conventions within which they operated. If the ambitions of rulers were the key element in causing wars, these frequently focused on concepts of honor (then a legal, not a moral, category) and glory. Dynastic concerns did not exclude other interests, but they remained a central feature of international relations. Not all monarchs sought participation in war, but most engaged in it at some time. This tendency was increased by a demographic profile that often led to young men succeeding to thrones, although some rulers, such as the Ottoman Sultan, Süleyman the Magnificent (r. 1520–66), were aggressive until their death. His willingness to take heavy casualties proved important to his capture of Rhodes in 1522.

The pursuit of dynastic claims could involve a large measure of opportunism. Indeed, in 1741, at a time when several claims were being advanced to the Habsburg (Austrian) succession, Robert Vyner referred critically in the British House of Commons to "one of those imaginary titles, which ambition may always find to the dominions of another." The fraudulent manufacture of some pretensions and the willingness of rulers to barter claims for "equivalents," other benefits to which they did not have a legal right, lends some substance to this charge. Thus, the allocation of territories during the Italian Wars of 1494–1559 was motivated by much besides dynastic legitimacy.

Consistency, of course, is a privilege of utopians. To note inconsistency, however, does not take away the need to assess the situation. Some monarchs and ministers, for example Cardinal Richelieu, the key French policy maker from 1624 to 1641, have been presented as adopting pretensions to serve their long-term plans of using dynasticism for the sake of *staatspolitik*. Much of the debate surrounding the policy of particular monarchs has revolved around this analytical device. The issue can be reconsidered by considering how far prudential considerations could affect the extent to which these claims were pushed. Richelieu was also criticized for pursuing *raison d'état* at the expense of religious consistency, specifically the interest of Catholicism. Under Richelieu, France fought Protestants at home and Catholics abroad.

Dynasticism was not a uniform practice. Understood as a situation in which rulers saw territories primarily as extensions of their own and their families' rights, interests, and will, and were determined to maintain and enhance this inheritance, dynasticism was far more common than if understood in terms of the pursuit of inheritance claims that had at least some possible basis in legality.

There was also a more functional aspect of rulership. The ruler was the war-leader, the head of the sociopolitical hierarchy and of the related socioeconomic system based on property rights that were in large part predicated on war or, at least, on military prowess. Ineffective military leaders lost crowns and eventually their lives, for example Edward II of England (1327),

Henry VI of England (1461 and 1471), and the Chongzhen Emperor, the last Ming Emperor of China (1644). The importance of conflict to position, political, social, and economic, was underlined by the extent to which warrior lineages, such as the *zamindars* of India and the *junkers* of Prussia, ruled the peasantry across much of the world, helping to ensure that local disputes were often conducted as feuds.

The importance of military prowess for honor, status, and wealth was highly significant to the way in which fighting could take place. Concern about the anonymous potency of the collective mass, such as pikemen and musketeers, reflected this importance. Nevertheless, traditional expectations and patterns of behavior adapted to the new emphasis on gunpowder weaponry. Aristocracies did not give up war because of the gun and the personal and social threats it posed. Instead, they retained their desire for mounted prominence, both in the cavalry and in positions of control over the infantry. Cavalry were not crucial around the world, but heroism was. In Japan, glory was gained by individual heroism in battle, producing a cult of the warrior, and heroism was assessed by an inspection of decapitated heads after battles, with warriors rewarded accordingly. So also with scalps in Native American warfare. Achievements were recorded in war tales in Japan as elsewhere. In Russia, promotion to boyar rank was dependent on prowess in battle and, alongside other aspects of serving the ruler, military service remained significant to status in the Table of Ranks introduced under Peter the Great (r. 1689–1725).[4]

A bellicose culture, society, and politics can explain the frequency of wars. That, however, is not the same as accounting for the outbreak of particular wars. There is a link in that the bellicosity encouraged the vigorous and violent pursuit of interests, claims, and disputes, whatever their origins and nature. Thus, the culture of power, politics, diplomacy, and warfare was central to their content. This remains the case today, even if the definitions and contexts vary.

Probing the links, at the level of individuals or ruling groups, between culture and the decision to fight at specific moments is possible but problematic. If the relevant material is often suggestive rather than definitive, it applies primarily to the culture and politics of power. However, that is the key level at which decisions were and are taken. Strategy and policy are intertwined and are framed and driven by these cultural elements. To treat the causes of war as a separate category, one essentially for "rational" strategic analysis, is a wish fulfillment rather than a description of practice, although it characterizes much of the literature.

An instructive attempt to employ the psychological approach was offered by E. H. Dickerman's work, which was based in part on medical records. He argued that the personal, especially sexual, problems and perceived inadequacies of Henry IV of France (r. 1589–1610) in the late 1600s led him to take an aggressive stance in the Jülich-Clèves succession dispute.[5] This resulted in 1610 in preparations for a major war with Spain that was only preempted by Henry's assassination by a Catholic "fanatic," in other words an individual who put religious views above secular loyalty.

The psychological approach is also valuable when considering the Japanese invasion of Korea in 1592. This invasion can be traced to the character and views of Toyotomi Hideyoshi, the unifier of Japan, who planned to conquer China and much else. It is possible to present a "rational," resource-based, account and, in particular, to argue that he wanted to gain fresh lands in order to provide for his warriors and keep them busy. Yet, it is more plausible to look at the extent to which his role and his self-regard were linked, with both dependent on continued warfare. Thus, the invasion of Korea came in a sequence of campaigning. Continual success, moreover, had led him to lose a sense of limits, while, anyway, the cult of the warrior discouraged an interest in limits. It was not so much a case of the misperception of realities as of no perception. In other words, an assessment of the capability of others was less relevant than a decision that such perception was not appropriate. "Overstretch" was not a useful concept.

Hideyoshi, however, had exceeded his grasp in large part because of Chinese intervention on behalf of Korea. That expanded conflict created a very different war, one essentially motivated by the Chinese sense of the integrity of a Chinese-dominated world order.[6] The parallel with Chinese intervention in "the" Korean War in 1950–53 is instructive. The ideologies deployed on both sides were different, but the dynamics of ambition and fear were similar. So also was the extent to which the resort to conflict appeared normative.

Any emphasis on monarchs, glory, and dynasticism has the advantage that it matches contemporary views. Such an emphasis, however, did not and does not lend itself to reifications of the state. Furthermore, the argument that the early-modern period witnessed the origins and growth of the modern, impersonal state, and that the causes of war should be assessed accordingly, can be queried. This is so on the grounds that insufficient evidence has been advanced to support the thesis, that much of the evidence relates to the writings of a small group of, arguably, unrepresentative thinkers, and that the political *practice* of the age was still essentially monarchical and in a traditional fashion. As a consequence, both the goals and the practices of rulership and foreign policy contributed to the frequency of confrontation and war. Violence was normal to these men, and they readily fought other expanding powers, more vulnerable states, and domestic opponents.

These considerations played a role in military methods. Rulers sought to secure triumphs that were as much symbolic as real, especially sieges that culminated in stage-managed surrenders and ceremonial entries, such as those of the Christian monarchs of Spain (Ferdinand of Aragon and Isabella of Castile) into Granada in 1492, which brought the *Reconquista* to a close.[7] So also for their grandson, the Emperor Charles V, Charles I of Spain. His reputation played a major role in the choices he made, not least in resolving the conflicting priorities of his various dominions. Far from being a mere chivalric fantasy, this concern with reputation was "the keystone in a conceptual arch forming the grand strategy that guided Charles and his advisers." Charles risked his own person in defense of his honor and reputation, which both increased the cost of his wars and made it easier to elicit aristocratic participation in his campaigning.[8] As protector of the Church against Protestants and Muslims, and its propagator in "the New World," Charles appeared to represent a new caesaropapalism that was designed to defend and revive the empire's sacred mission. Far from suggesting limits, Charles carried Habsburg power and the Crown of Spain to hitherto unsurpassed and unimagined heights, creating, and linking, two types of empire. The grandiloquent Latin inscription above the entrance to the royal mausoleum in the Escorial palace built near Madrid by his son, Philip II of Spain, was to refer to Charles as "the most exalted of all Caesars."

So also for others. Heroic conceptions in Europe could lead to an emphasis on hand-to-hand conflict and on officership in the more socially distinguished cavalry. The nature of exemplary military conduct, however, was very different in China. The social and cultural contexts there led to far less of an emphasis on military service than in Christendom or Japan. This was seen with the lack of such activity by most Chinese Ming emperors and the extent to which senior advisers had no experience of such service. Many were eunuchs.

The role of ministers might appear an important qualification of any argument centered on monarchs. Thus, Sokollu Mehmed Pasha, Ottoman Grand Vizier from 1565 to 1579, encouraged caution and peace. However, the position of ministers totally depended on royal favor. Indeed, although the Ottomans ended their conflicts with Venice and Spain, they began, in 1578, a major war with Persia. Moreover, dynastic considerations clearly played an important part in ministerial correspondence and discussions.

CIVIL WAR

As already noted, civil war is in part a matter of definition. That was certainly the case with the period, with not only rebellions but also wars of unification and consolidation. The latter involved the suppression of opposition that was defined as rebellion, but that could also display inchoate statehood, or that might lay claim to it.

The sovereignty of the ruler in individual polities did not preclude traditions that bad kingship might be redressed by rebellion. In doing so, foreign assistance could be summoned. A good example was provided by the outbreak of the Thirty Years' War. In 1618, the Bohemians rejected the traditional authority of the Austrian Habsburgs and followed this up by electing a different monarch, which led to conflict. This process can be seen as rebellion, war, and both. Bohemia was an elective monarchy, but successive Habsburgs had held the crown for nearly a century. The Bohemians, instead, elected a Calvinist, Frederick V of the Palatinate, while the Habsburg claimant, the Catholic Emperor Ferdinand II, sought, and received, military assistance from Bavaria, Saxony, and Spain, and won control of Bohemia as a result of victory over Frederick at the battle of the White Mountain (1620).

Rebellion leading to an acknowledged and independent state with a sovereign right to wage war was rarely successful, although the Swiss Confederation and the United Provinces (the Dutch) gained such a status and the United States was to do so in the eighteenth century. It is, therefore, difficult to decide when it is reasonable to cease talking of rebellion, as in the Dutch Revolt that began in 1566–68, and, instead, begin considering it as a war, as in the "Eighty Years' War" between Spain and the Dutch, which, after a truce in 1609–21, finally ended in 1648. The key element in the Dutch Revolt was not the riots in 1566–67, but rather the dispatch of a powerful Spanish army that brutally imposed control. Although initial opposition was crushed in 1568, from 1572 there was always part of the Low Countries outside Spanish control.

Spain's eventual failure serves as a reminder of the interrelationship of conflict and politics and the need to move from outputs to an outcome. The Dutch were far from typical. Early-modern Europe is littered with instances of polities that failed to become independent states, not only Bohemia, but, for example, Catalonia in 1640–52, Hungary in 1704–11, and Ukraine in 1708–9. So also with India, for example with Bengal, where the nominal sovereignty of the Mughals was repudiated in 1574 or, somewhat differently, in southwest China, where Yang Yinglong's drive for autonomy was crushed in the late 1590s.

Rebellions could encourage external intervention and thus trigger wider conflicts. In the French Wars of Religion that began in the 1560s, there was notable foreign intervention by England, Spain, and German princes, and England intervened when they resumed in the 1620s. Like many other wars, the French Wars of Religion can be seen as individual struggles, separated by truces and peace treaties, each with particular causes. Nevertheless, there was also a common strand of causation. The root causes of dissension in France remained, notably religious division, aristocratic factionalism, and royal weakness. Aristocrats and churches created military forces.[9] Moreover, whatever their length, periods of peace were characterized by a degree of mutual suspicion that helped fan the flames when conflict resumed.

There was a genuine popular dimension to the wars: they were not simply a matter of the struggles of king, nobles, and their affinities. If the Thirty Years' War was one example, the Scottish rebellion of 1637 was another. Due to Charles I's total failure to suppress it in the First (1639) and Second (1640) Bishops' Wars, the Scottish rebellion became a British-wide civil war, eventually embracing England, Ireland, Wales, and England's overseas colonies, lasting until 1652. The relatively limited goals of 1637 became far more radical, which serves as a reminder of the difficulties of treating a war as a unit: the causes that were in dispute in 1648 and 1650 were very different to those in dispute in 1637 and 1642, although there was a linkage. However, discussing the cause of a war can make the later stages appear far more inevitable than was the case, and can also ignore the extent to which new combatants and alignments came forth, and frequently in a highly unpredictable fashion. The extent of likely "foreign" intervention was unclear, as was its eventual success. Indeed, there was a major contrast between Charles I's failed attempt to defeat Scottish opposition and the far more successful attempt made to Oliver Cromwell from 1650.

The frequency of rebellions across the world does not mean that they took place in a similar context. In the Islamic world and China, the rulers were able to benefit from greater authority than their Christian counterparts. The latter compromised far more with existing interests, and ruled by virtue of, and through systems of privilege, rights, and law that protected others as well as themselves.

CONCLUSIONS

War was discussed as an aspect of legitimate power, and therefore of sovereignty, both secular and religious, insofar as the two were separable. This, however, left unclear how peace was to be preserved in the absence of a dominant sovereign ruler and power, such as China. In Christendom, the alliance of Emperor and Pope had offered such a means, but this alliance, already highly tenuous, lost authority and power in this period. As a consequence, the balance of power, or a nonmechanistic equivalent idea, was pushed to the fore. In practice, however, the idea, which itself served specific purposes,[10] did not readily operate to lessen warfare, as the seventeenth and early eighteenth centuries were to show.

NOTES

1. Taddesse Tamrat, *Church and State in Ethiopia, 1270–1527* (Oxford: Oxford University Press, 1972), 294–300.

2. Richard Pankhurst, *The Ethiopian Borderlands* (Lawrenceville, NJ: Red Sea Press, 1997), 222–23.

3. Jonathan Spence and John Wills, eds., *From Ming to Ch'ing: Conquest, Region and Continuity in Seventeenth Century China* (New Haven, CT: Yale University Press, 1979); Frederic Wakeman, *The Great Enterprise: The Manchu Reconstruction of Imperial Order in Seventeenth-Century China* (Berkeley: University of California Press, 1985).

4. Brenda Meehan-Waters, *Autocracy and Aristocracy: The Russian Service Elite of 1730* (New Brunswick, NJ: Rutgers University Press, 1982).

5. E. H. Dickerman, "Henry IV and the Juliers-Cleves Crisis: The Psychohistorical Aspects," *French Historical Studies* 8 (1974): 626–53, and "Henry IV of France, the Duel and the Battle Within," *Societas* 3 (1973): 207–20; Dickerman and A. M. Walker, "The Choice of Hercules: Henry IV as Hero," *Historical Journal* 39 (1996): 315–37.

6. Kenneth Swope, *A Dragon's Head and a Serpent's Tail: Ming China and the First Great East Asian War, 1592–1598* (Norman: University of Oklahoma Press, 2009).

7. Larry Silver, *Marketing Maximilian: The Visual Ideology of a Holy Roman Emperor* (Princeton, NJ: Princeton University Press, 2008).

8. James Tracy, *Emperor Charles V, Impresario of War: Campaign Strategy, International Finance, and Domestic Politics* (Cambridge: Cambridge University Press, 2002), 38.

9. S. Carroll, *Noble Power during the French Wars of Religion: The Guise Affinity and the Catholic Cause in Normandy* (Cambridge: Cambridge University Press, 1998).

10. Benno Teschke, "Revisiting the 'War-Makes-States' Thesis: War, Taxation and Social Property Relations in Early Modern Europe," in *War, the State and International Law in Seventeenth-Century Europe*, ed. Olaf Asbach and Peter Schröder (Farnham: Ashgate, 2010), 59.

Chapter Four

Limited War? 1650–1790

The customary chronology in the discussion of war and its causes was that of a decline in ideological factors between the "Wars of Religion" and the outbreak of the French Revolutionary Wars in 1792. Bellicosity, nevertheless, remained a key factor. However, this is very much an agenda that is set by Western concerns and developments in that it focuses on the Western interest in Western history. In addition, there is a teleological focus on a state system supposedly created by the "Westphalian Settlement," the Peace of Westphalia of 1648 that brought to an end the Thirty Years' War. That settlement is commonly presented as a triumph of reason and restraint, in the shape of an agreement to operate an international system based on the mutual respect of sovereign powers and, in particular, an agreement to accept confessional plurality, at least in the form of different types of Christianity, as sole state religions. To contemporaries, the latter was probably more significant than subsequent secular commentators have suggested with their focus on a system of mutual respect, which is generally presented largely, or exclusively, in lay terms.

Whether respect and restraint are emphasized in secular or in spiritual terms, the reality, however, was that the Peace of Westphalia did not usher in a new age. Indeed, attempts to advance secular and confessional interests continued in a manner that was far from limited. The Westphalian settlement also meant nothing across most of the world.

BELLICOSITY, CULTURE, AND RULERSHIP

International rivalry was one of force and power, with warfare, in Europe and more generally, linked to state-forms, governmental development, and the culture of majesty, notably royal ambition. Around the world, but to a varying extent, tensions between tribal and state forms of government were significant, although, alongside contrasts between the two, there were also overlaps.[1] Across the world, *gloire*, loosely translated as the pursuit of glory, was important in causing and sustaining conflict, in setting military objectives, and in celebrating success. This reflected the extent to which the notion of "the state" as a rational contrast to the monarch had more meaning in political theory than in practice. Military success brought exemplary purpose and fame, endorsed legitimacy and dynastic/state/national exceptionalism, and acted as a lubricant of obedience in crown-elite relations. Military heroism played a major role in the representation of states,[2] and this role affected literature, history, and the arts.[3] Indeed, in his 1792 treatise, *Yuzhi shiquan ji* [*In Commemoration of the Ten Complete Military Victories*], the Qianlong emperor of China (r. 1736–96) reached ten by including failures. The points made in this paragraph could all have been made for previous centuries, which indicates the degree to which there was no real break, either in the mid-seventeenth century or at the close of the fifteenth.

War offered the possibility of strengthening the dynastic position, both domestically and internationally, and also of enhancing territorial control. This element was important to major rulers pursuing large-scale goals, but also to lesser princes, such as Max Emmanuel, Elector of Bavaria during the War of the Spanish Succession (1701–14), seeking to sell his participation in the conflict for gains in territory and status.[4] This was the buffer state or player as agent, a situation that continues to the present. In the event, Max Emmanuel's support for Louis XIV of France (r. 1643–1715) led to defeat and the conquest by Austria of his territories, which he only regained as part of the peace settlement.

A key element of Louis's bellicosity was that he had little realistic sense of how the wars he began would develop diplomatically, militarily, or politically, but then that has been true across history, as seen with the Iraqi invasions of Iran and Kuwait in 1980 and 1990 respectively and the American-led invasion of Iraq in 2003. Thus, in 1672, rather than fighting a limited war with the Dutch, who, after being invaded, were willing to offer terms, an overconfident Louis, hopeful that the war would widen to include Spain, so that he could resume the conquest of the Spanish Netherlands (Belgium), issued excessive demands, including for major territorial gains and the acceptance of Catholic worship. The last reflected the extent to which religion was an important goal for all the powers, with Louis also determined to gain prestige by being seen as a champion of Catholicism, supplanting the Habsburgs in this role. In 1673–74, the conflict changed shape and broadened, as Austria, the Dutch, Spain, and Lorraine agreed in 1673 to force Louis back to his 1659 frontiers. Louis was happy to see Spain enter the war, but not Austria, nor, in 1674, to lose the alliance in England.

Similarly, Louis miscalculated in 1688 when he assumed that his use of his army to enforce his interests in the Rhineland, notably in Cologne and the Palatinate, would lead to a limited, short, and successful conflict in which German rulers would desert Leopold I of Austria, who would be driven to terms by French triumphs. Instead, the struggle broadened and Louis found himself facing a powerful coalition. So also with the miscalculations of the rulers of Denmark, Russia, and Saxony-Poland who, in 1700, launched what became the Great Northern War in an attempt to partition the Swedish overseas empire. Dying in 1715, Louis XIV told his successor, Louis XV, "Above all, remain at peace with your neighbours. I loved war too much."

Military command could be separated from rule, as was demonstrated by the great success and considerable importance of Prince Eugene, John, 1st Duke of Marlborough, and Marshal Saxe, leading generals for Austria, Britain, and France respectively. In addition, George Washington and the Marquis of Lafayette, both heroes of the American Revolution, were to offer a model of modern military celebrity in which political dedication to a cause they did not control was a key aspect of their appeal.[5] In the Turkish empire, military leadership on campaign was increasingly by Grand Viziers (leading ministers), and not by sultans, a pattern that had begun after Süleyman the Magnificent died in 1566: Ahmed III accompanied the army in 1715 and 1717, but remained far behind the front line, and in 1730 did not even set out on campaign, helping precipitate the rebellion that led to his overthrow.

Nevertheless, military command was generally a key aspect of rulership, a situation encouraged by the view that waging a lawful war required a declaration of war, and that only true sovereigns could declare war.[6] Rulers earned kudos from their military prowess, for example George II of Britain as a result of victory at Dettingen in 1743. John Campbell's comment on Frederick William I of Prussia (r. 1713–40)—"he made his troops his delight, and led all his days rather a military than a court life"—could have been repeated for other rulers. The wearing of uniform was important for monarchs in some Western states.[7] So also were military reviews. Thus, in 1777, Ferdinand I of Naples spent much time exercising his troops, including taking part in a mock campaign he planned that entailed constructing a camp and staging mock battles and a siege.[8]

Moreover, as for most of history, and indeed still for some states today, many of the commanders of the period were also rulers. Most, such as Peter the Great (I) of Russia (r. 1689–1725), Frederick the Great (II) of Prussia (r. 1740–86), and the Kangxi emperor of China (r. 1661–1722), inherited the right to rule, and then used war to enhance their assets. Peter, in particular, transformed both army and state in order to increase Russian military effectiveness and win territory and prestige. He inherited his first war with the Turks, but deliberately started a second in 1711. As one of the allied rulers plotting to deprive Sweden of its overseas empire, Peter also played a role in

beginning the Great Northern War (1700–21) and persisted in this struggle
longer than his allies. He was very much a ruler who fought, taking part in
major battles, notably the crucial victory over the Swedes at Poltava in 1709
and the defeat by the Turks at the Pruth in 1711. The Great Northern War
over, Peter led an invasion of Persia in 1722, an invasion that produced
territorial gains only to lead, after his death, to a withdrawal that reflected an
over-extension of power in the face of a resurgent Persia.

In contrast to inheriting power, Nadir Shah of Persia, having become
crucial, as a commander, to military success against the invading Turks, took
over a failing empire in the 1730s, declaring himself Shah in 1736, and gave
Persia a military dynamism until his assassination in 1747. Even more than
Peter, he campaigned widely, leading armies against the Turks, but also, in
1739, into northern India. This was war of a type familiar to the Mongols
under Chinggis Khan (c. 1162–1227) and many others. Given this continuity,
it is interesting to consider why any particular psychological or practical
transformations in the causes of war should be anticipated in the far smaller
period of time since then.

War and military activity played important roles in a culture of power to
which honor and prestige were significant, let alone such related issues as
insecure and provoked masculinities.[9] Concern for personal honor was particularly apparent in the case of officers, and greatly affected command practices and issues.[10] The effects, however, varied. No Western ruler matched
'Alaungpaya of Burma (r. 1752–60), who judged men largely by the number
of heads they could produce after a battle. Moreover, unsuccessful Burmese
officers were executed—a practice also seen with the enforced suicides of
commanders in China. With the prominent exception of Admiral Byng, shot
in 1757 after a court-martial for failing to relieve the French-besieged British
garrison on the Mediterranean island of Minorca the previous year, the exe-
cution of commanders was uncommon in the West until Revolutionary
France, from 1792, treated failure as a demonstration of treason. The
American revolutionaries had not followed this course, which was a prime
instance of the less radical path followed there than in the case of France.

Again on a long-standing pattern, the cultural dimension extended to the
aesthetics of warfare, with rulers placing particular value on a good-looking
army. This preference played a role in recruitment, encouraging the acceptance of tall men over short ones, notably by Frederick William I of Prussia.
For the same reason, many Western rulers designed their soldiers' uniforms,
with an emphasis on having the army look good as well as seeking to assist
local textile industries.

Military activity also fulfilled narratives and models of imperial and royal, destiny and role. The repute of rulers and the fame of ancestors were
echoed.[11] This can be very much seen in the iconography employed by Louis
XIV in his new palace at Versailles, a palace that others in Europe sought to

emulate as in Berlin (Prussia) and Het Loo (the Netherlands). The ruler as warrior was the key theme, and one intended to enable Louis to outshine his predecessors and contemporaries, and to set a model for his successors.

The willingness of rulers, commanders, and combatants not only to kill large numbers, but also to accept heavy casualties, was an important feature of military culture, although the latter acceptance varied. Preserving the army was the priority, but there was a greater willingness to take casualties than with much, although by no means all, modern warfare, and certainly insofar as regular forces are concerned. Based on the cost of training modern troops and their relative rarity, a functional explanation of this current unwillingness can be advanced, one that contrasts it with the situation in the eighteenth century. However, social, cultural, and ideological factors are, and were, more significant, not least of which is the contrast between modern individualism and hedonism, both of which discourage an acceptance of casualties (even though the population is far larger), and, on the other hand, earlier concepts of duty and fatalism in a much harsher working environment. The extent to which this sense of duty and fatalism can be related to levels of habitual violence in society is unclear.[12]

The acceptance of casualties was crucial to the bellicosity of the age. On the pattern of earlier moralists, principally religious writers, Enlightenment individuals in Christian Europe and North America might criticize all, or much, of this belligerence, presenting it as a pointless and indeed dishonorable bloodlust. However, these views had scant impact on the goals and means of waging conflict. Instead, the continued normative character of resorting to warfare was more notable, and across the world. Rulers, ministers, and commanders, very much an overlapping group, tended to see their wars more as transitioning into each other than as distinct conflicts punctuated by periods of distinct peace. In part, this was an objective response to particular bilateral relations, for example between China and the Zunghars, the Mughals and the Marathas, Russia and the Turks, or Britain and France. There was also the contrast between the concept of a just and lasting peace and the belief that this was likely to be illusory due to the failings of human society. War thus was normative.

Wars were believed to be not only necessary, but also, in at least some respects, desirable. In each respect, they could be presented as just. This conviction proved a key context for the eighteenth century, and also for those that preceded and followed it, again undermining the idea of significant changes through time in the longer eighteenth century. The belief in necessity also helped explain the attitude in combat zones to those who would later be called civilians. This issue included the response to conventions that restricted violence, at least within cultural zones, such as Christendom and the world of Islam.[13] In practice, these conventions were frequently ignored.

The nature of rule was a central element in the military history of the period. Military systems with political continuity and stability, and administrative strength, particularly China, France, Britain, and Russia, proved more able to sustain a projection of their power than monarchies on horseback, such as those of the Zunghars, Nadir Shah in Persia, Ahmad Shah Durrani in Afghanistan, and Napoleon, or, with the exception of their not using horses, 'Alaungpaya in Burma, and Tamsin in Siam (Thailand). Ultimately, therefore, military history is an aspect of the other histories of the period, as well as contributing greatly to them.

Environmental history was one such. The dependence of operations on the weather and climate was a key element, as "the present scarcity of corn and all other provisions"[14] could delay or prevent moves, or be believed likely to do so. Harsh winters and springs delayed the appearance of grass, affecting the moves of cavalry. Indeed, insofar as war could be a series of campaigns, each bounded by the weather, then the decision to start anew could be influenced by the latter. Nevertheless, alongside potent limitations, their potency in part due to their unpredictability, governmental development was significant to military capability, notably in helping produce, deploy, and sustain the necessary resources for war,[15] but, moreover, as an aspect of a system of rule that provided the political stability valuable for military activity.

Aside from serious issues in the coverage of warfare in this (and other) centuries, there are concerns about standard explanations. As a reminder, for example, of the need for care in drawing clear causal conclusions in the sphere of international conflict, and therefore suggesting a willingness to fight based on an assumption of success, governmental systems of continuity, stability, and strength were not invariably successful in war. The Chinese discovered this in Burma and Vietnam, the Russians in Persia, the Afghans in Persia and the Punjab, and the British, eventually, but more centrally to their military effort than in the Chinese and Russian examples, in North America. In each case, both conflicts and failures can be regarded as those of imperial overstretch, albeit in very different contexts, the last example, the American Revolution (1775–83), having an element of civil war not seen in the other examples cited in the previous sentence. However, although the thesis of such an overstretch, in aspiration and attainment, appears seductively clear, its application is not readily inherent to particular circumstances. Instead, far from being readily obvious, overstretch emerged through the warfare of the period, warfare that helped define imperial limits and, more generally, mold, as well as register, the politics of the period.

A common requirement of the ruler-leaders in this era was the demonstration of both political and military skill in order to win *gloire*, with the military dimension closely associated with the political. As with other periods of conflict, wars were best fought sequentially, not simultaneously. It was necessary to divide opponents, to create tensions in their relationships, indeed alliances, and to fight them in sequence. This was a practice at which Frederick the Great was generally adept as, in more propitious circumstances given his far greater power, was the Qianlong emperor of China: war with Burma followed that with the Zunghars, and war with Nepal that with Vietnam.

Correspondingly, Louis XIV proved unable to move from a successful opening of wars to an ability to overcome the opposing coalitions to which his conflicts gave rise. With Louis, as with other cases, for example World War II, it is important to see two sets of "causes": those of the initial conflict and those that explain subsequent entries into the war. With Louis, there was a growing concern about his intentions, and understandably so given his disinclination to accept limits. This concern can seem a quasi-automatic response to the threat of a new hegemony, or, conversely, a reluctance to resist until success seems assured. Moving from the model of hegemonic drive and resulting resistance, it is also appropriate to note the glory involved in resisting Louis. This was very much seen with the image of William III of Orange and Victor Amadeus II of Savoy-Piedmont. It was also important, for both Louis and his opponents, to hold together constituencies of interest, be they international alliances or the groups within countries backing a war effort. Doing so made the resort to war far more attractive.

The political dimension to conflict is also captured in the stress in the literature on state-building, but there is a danger that that dimension of causes and consequences is approached in an overly schematic and "modern" fashion, and without giving due weight to other political aspects of warfare and the related military aspects of politics. The notion that war makes states and states make war is seductively clear, but also begs a lot of questions about both the processes involved and the extent to which war also fulfills other purposes, as well as frequently undermining states as a result of the burdens incurred, whether resource or political, structural or contingent.

In addition, it is important to emphasize the degree to which rulers, in making war, called upon non-bureaucratic processes to raise and support forces. These processes had significant consequences for the way in which war was pursued. Moreover, the extent to which states make war was (and is) not automatic, consistent, or similar in frequency. Rather than simply viewing this question in terms of a model, whether of war, the international system, or of state-building, a model that takes precedence over particular circumstances and individual decisions, it is also pertinent to focus on these circumstances and decisions.

These circumstances and discussions are of particular significance when considering choices made between possible challenges, for these choices indicate that geopolitical and other factors also play through specific issues. For example, by turning east against Afghanistan and India in the late 1730s, each a different military environment, Nadir Shah reduced the pressure on his former opponent, the Turks. By doing so, he enabled the latter to concentrate on war with Russia and, more successfully, Austria, and to great effect in 1739 when the Austrians, as a result, lost Belgrade. In contrast, Russia and Austria had each benefited from the end in 1735 of the War of the Polish Succession, which had reduced the pressures stemming from real or likely war on their western fronts.

Governments could also be divided over policy, as the Chinese one was over going to war with the Zunghars in the 1750s, the British over war with Spain in 1729, 1738–39, and 1761, and with Russia in 1717–20 and 1791, the French over war with Britain in 1770, 1778, and 1790, and the Americans over policy toward Britain and France in the late 1790s. Such division interacted with more general and more specific cultural pressures for bellicosity.

The role of ideology also has to be assessed. If religious differences played a part in rivalry between Shi'ite Persia and Sunni Turkey, they did not prevent peace between the two for most of the second half of the century, although they fought in the Caucasus even when not elsewhere. In the Punjab in northwest India, religious hostility between Sikhs and Afghans led, in the 1760s, to sustained and eventually successful opposition to the latter. Subsequent Afghan invasions of northwest India saw much violence directed against Sikh practices. Tension over the radical plans of the French revolutionaries may have led to war with Austria and Prussia in 1792, but Britain, the Dutch, and Spain did not enter the war with France until 1793, while Prussia and Spain left it in 1795 and Spain swiftly allied with France. In the early 1790s, Russia was more concerned about controlling Poland than opposing French radicalism, and crushed Polish opposition accordingly in 1794.

Far from there being any fixed relationship between war and politics, it is the flexible nature of the links that helps explain the importance of each to the other. Military activity certainly altered the contours and parameters of the politics that helped cause it, and sometimes of the states involved in conflict. In some cases, military activity had a comparable impact on social structures. The centrality of war as a basis of change, however, does not mean that there was a consistent pattern of cause or effect.

WARS ACROSS CULTURES

As before, there were important questions of definition when assessing wars across cultures. Although their ruling ideology was generally different, many major states were clearly hybrids, including Mughal India, Safavid Persia, Ottoman Turkey, and Manchu China. Indeed, British India was to come into this pattern. The culture, political system, ideology, and military arrangements and methods reflected the traditions of conquerors and conquered. In part, this was a matter of assimilation and in part of the maintenance of separate practices. When the Mughals campaigned against the Uzbeks across the Hindu Kush in the 1640s or the Manchu conquered Mongolia in the 1690s, it is difficult to see such campaigns in terms of the clash of totally different cultures. Nevertheless, there had been a process of assimilation on the part of conquering Manchus, Mughals, Safavids, and Ottomans, and, partly as a result, the element of cultural clash was apparent. This can be seen in the Afghan overthrow of Safavid Persia in the 1720s.

The Manchu fought to expand. Their expansionism was mostly imperial, for glory and possessions rather than for resources and trade. In contrast, the earlier Ming dynasty had fought mostly to preserve itself and its dependents, such as Korea. Manchu warfare was certainly not limited, any more than that of the Europeans against Native opponents in North America or the Russians against opponents in Siberia.

Yet, there could also be cooperation. Like the Ming, the Manchu sought the support of particular Mongol tribes, and, thus, to lessen the Mongol threat. It was also the case with European relations with Native Americans in North America. The fur trade led to the creation of political-economic networks and to the involvement of Europeans in Native wars and vice versa, a process facilitated by, and in part due to, the increasing provision of European arms and ammunition. However, this involvement was not a case of more of the same. In North America, as elsewhere, European intrusion intensified war among non-state peoples.

The notion of hostility across cultures has to be qualified by a realization of the degree to which intracultural wars were as important or could take precedence. In the first half of the eighteenth century, the Russians devoted more effort to fighting Christian European neighbors, especially a life-and-death struggle with Charles XII of Sweden, than to fighting other polities, none of which posed a comparable threat, although they did devote much effort to war with the Turks in 1711 and 1736–39. In North America, the British, French, and Spaniards focused their efforts on conflict with each other, rather than with the Native Americans, and this is even more apparent if expenditure on fortifications and naval support is also considered.

In India, the British were more concerned about France in the late 1740s than about native states. In the Carnatic, this situation changed only after the defeat of the French in 1760–61: the battle of Wandewash and the fall of Pondicherry were crucial preludes to more assertive British policies toward native rulers, notably Mysore and the Marathas. The destruction in India of an Anglo-French balance and of a French threat to British interests were followed not by peaceful hegemony in regions where the British were strong but, instead, by bouts of expansionist activity.

In India, and more generally, there was no necessary pattern of continual warfare. Instead, alongside cultural and other factors that encouraged expansionism, it is more helpful to consider peace, wars, and the cause of the move from one to the other. In part, this was the case, as with all wars, of the complex relationship of fear and opportunity. Native Americans could attack European settlers because they feared their advance and despaired at the inability or unwillingness of colonial authorities to control it. This happened in the Carolinas in 1715, in what was to be Vermont in the 1720s, and in Pontiac's War in 1763–64, a war presented by European Americans as a rebellion or rising.

Fear also played a role in Asia. The Chinese feared the creation of a hostile Mongol confederation from the 1690s and notably the consequences in Tibet as well as Mongolia. Fear was seen elsewhere. Burmese conquest in 1767 made later Siamese (Thai) rulers fearful, encouraging them to take an aggressive stance and also to dominate the lands through which the Burmese had advanced, and might do so again.

Religious tensions were also significant in several Asian struggles. In western Sichuan in China, the First Jinchuan War of 1747–49 broke out as the emperor sought to bring the essentially autonomous Gyalrong or Golden Stream tribes under administrative control. Religious animosity played a role, as the Golden Stream followed the indigenous, animist, Tibetan Bon religion and Tibetan Buddhism's Red Hat sect, resisting the Yellow Hat sect that the emperor supported. In part, the conflict therefore involved a struggle between different types of prestige, magic, and providential support. This element was more generally true in East Asia, where leaders were sacral figures.[16] Similarly, in Persia, Nadir Shah's wish to resolve the schism within Islam and to integrate Shi'ism into Sunni'ism helped encourage opposition to him in Persia as well as the animosity of Turkey, whose ruler claimed religious authority as Caliph.

Religion was a significant element elsewhere. In Spain, the tradition of conquest for Christ remained strong and difficulties experienced in converting natives led to a harsh attitude that also justified imperial expansion. When Nojpeten, the capital of the Maya people known as Itzas, was stormed in 1697, Martín de Ursúa, the interim governor of Yucatán, ordered his men to plant the flag with the royal arms of Spain and religious standards among the Itza temples "in which the majesty of God had been offended by idolatries." Ursúa thanked God for his victory and then joined soldiers and Franciscans in destroying a large number of "idols." If gold or silver, such idols were melted down. This was religious war against opponents presented as guilty of human sacrifice, cannibalism, and killing priests.[17] Five years earlier, when the town of Santa Fé was regained by the Spaniards after the Pueblo rebellion of 1680, there had also been a reimposition of Catholic control. Franciscan priests absolved natives of the apostasy and baptized those born after 1680, with the governor serving as godfather for the children. Religious identity remained crucial. Following the capture of Oran in 1732, Benjamin

Keene, the British envoy, reported: "there is scarce a Spaniard who does not think himself half way to his salvation by the merits of this conquest" from the Muslims. [18] Large-scale Spanish attempts to capture Algiers, mounted in 1775 and 1784, failed badly. Generally forgotten, they serve, however, as a reminder of the significance of religious struggle. Although it was to be France that conquered Algeria from 1830, there is a link from the *Reconquista* to the struggle in the sixteenth to eighteenth centuries to dominate the western Mediterranean, and then on to expansionism in Morocco from the mid-nineteenth century. The values of the army can in particular be located in this context.

In these and other circumstances, opportunity, as well as fear and anxiety, played a role, but it could be a case of opportunity rather than fear or anxiety. A perception of weakness providing the chance for great gain encouraged the Afghans to attack Persia in 1722 and Nadir Shah to invade Mughal India in 1739. Both attacks were successful.

EUROPE AND THE MEANS OF WARFARE

The degree to which the causes of war can be separated from their means invites discussion, for this as well as other periods. The role of change in the means is of particular interest. In 1815, two of the three leading military powers were European and all three—Britain, China, and Russia—were the same as the leading powers a century earlier. All the intervening fighting had had many consequences, but in the long term it had confirmed the verdict on the conflicts in 1690–1715 that had left Britain, China, and Russia the leading powers in this period, indeed considerably strengthening them. That, however, is neither the impression created in military history as a whole nor that which can be readily gauged from a consideration of European military history in this period with its concern with Frederick II and Napoleon. These observations are linked. There is a general tendency to note the significance of war in European history, but somehow to marginalize the discussion of warfare in the sense of the fighting. Instead, the relevant subdiscipline that has attracted attention is that of war and society. While interesting, it, however, offers little for the discussion of such topics as the relative effectiveness of military systems and the reasons why, in particular conflicts, some succeed and others do not.

This neglect encourages a focus on an easy account in military history that serves apparently to settle the topic, and that has implications for the discussion of the causes. The account is that of military revolutions, an approach that draws on Hegelian-Marxist assumptions about process as well as teleological views of, and about, change. Each characteristic can be found in the case of the two prime instances for European military history, first the

supposed military revolution linked to the transition to modernity at the close
of the Middle Ages, and, second, the supposed military revolution linked to
the concept of the people under arms, a concept applied to the American,
French, Haitian, Latin American, and other revolutions from 1775. This approach, however, is deeply flawed in its underlying assumptions and seriously wrong in specifics.

The established accounts focus on Western, not Eastern, Europe and offer
a core-to-periphery model of the diffusion of innovation, leadership, and
lessons. The key struggles apparently are those within the core. Thus, from
the onset in 1494 of another group of conflicts in Italy to the end of the
seventeenth century, the standard emphasis in on a struggle between Spain
and France. This baton-changing approach to military history moves on to
center on Prussia from 1740, only for France to return to the fore from 1792
with the victories of revolutionary forces.

This portrayal is taken further with Napoleon's successes, only for him to
fail totally in 1812–15 due to his egregious mistakes, military and political,
and others "catching up" with his innovations. Moreover, as further evidence
that France really "won," Napoleon's failure is explained partly in terms of
the greater resources of his opponents—an approach that prefigures one of
the standard (and deeply flawed) interpretations of German failure in World
War II.

The baton-changing approach has many problems, but, again, it proves
attractive because, in its clarity, it offers a high rating for that most seductive
of ratios: ease of interpretation in relation to the amount of work. In practice,
as consideration of any period will reveal, there was no paradigm state of
warfare, no single model of improvement. Instead, alongside borrowing,
there were autonomous developments within a general context of fitness for
purpose. Militaries must respond to tasks and goals that are specific to particular states and occasions, and must do so within particular social, economic,
and political parameters. In the last case, the political acceptability of a large
army was very different in Britain and Austria in the eighteenth century. The
causes of war can be similarly placed.

Turning to the eighteenth century, the key standard interpretation addresses a transition from *ancien régime* to revolutionary warfare. This interpretation employs concepts of modernity and modernization that are problematic
in military history and elsewhere, not least because employed in an essential-
ist and teleological fashion. The teleology for the eighteenth century sits
within a broader one that assumes an inevitable development toward nationalism, peoples' warfare, and the industrialization of war. Each of these is
presented as more effective than hitherto, with, in turn, even greater effectiveness derived from their interaction. From that perspective, an earlier age
appears anachronistic, its warfare limited, its militaries bound to fail or at
least be superseded, and its command practices and culture outmoded, unpro-

fessional, effete, and clearly ready for replacement, indeed sufficiently so for an understanding of this point to be a definition of competence.

Always problematic, that approach looks increasingly flawed in light of the substantial but varied changes in conflict since the Cold War ended. There were/are also misreadings of the eighteenth century. Questions of significance are frequent. There is the repeated problem of focusing on a particular military, for example the Prussian army under Frederick the Great, and treating this military as if it defines progress and therefore importance, while its victories, in turn, underline relative capability and thus demonstrate this progress. Aside from the serious conceptual and methodological problems posed by such circular arguments, there are also those posed by a disinclination to draw sufficient attention to the failures of those supposed paradigm powers. For example, Prussia was repeatedly unable to inflict significant defeats on Austria in 1761–62 or 1778–79, and the Prussian military system totally collapsed at the hands of Napoleon in 1806.

Moreover, this approach shortchanges other powers. Focusing on Prussia in mid-century leads to a downplaying of French successes in the Low Countries in 1744–48. Focusing on France in 1792–1809 leads to a downplaying of Russia's repeated successes in that period, successes that make more sense of Russia's subsequent ability to defeat France. So also with Britain's history of naval triumph and transoceanic power projection. British victory over France under both heads in 1793–1815 was not new but looked back on a continuity that reflected powerful institutional, doctrinal, and contextual strengths, as well as the grit of experience, morale, and sense of purpose.

In assessing changes, there is the repeated question of which changes to consider and how they might have affected the turn to war. Technological and tactical circumstances in European warfare were far more similar in 1715–1815 than they were to be over the subsequent century. Moreover, although the changes in 1615–1715 did not match those in 1815–1915, they were also greater than those seen in the French Revolutionary and Napoleonic wars. Indeed, considering the relationship between the causes of war and the nature of war, it is worth focusing on the significance of the move in 1690–1700 away from the musket-pike combination that had become dominant in Christian Europe in the sixteenth century. Turning to the latter, the particular combined armed tactics resulting from the supposed military revolution of the sixteenth and seventeenth centuries were not "new" in the sense that both cavalry and infantry, and, for each, missile and stabbing or thrusting weapons, had long been combined. In addition, as far as the sixteenth and seventeenth centuries were concerned, successful combination proved easier to discuss in training manuals, which emphasized drill, and to attempt in combat, than they were to execute successfully under the strain of battle. Furthermore, the contrasting fighting characteristics of the individual arms— muskets, pike, cavalry, and cannon—in the sixteenth and seventeenth centuries operated very differently in particular circumstances, which posed added problems for coordination. So also did the limited extent to which many generals and officers understood these characteristics and problems.

Issues of firepower and tactical coordination help explain the significance of the move in the 1690s away from the musket-pike combination to the fusion presented by firearms mounting bayonets, in which each foot soldier carried a similar weapon. The simultaneous shift from matchlock to flintlock muskets improved their reliability as ignition systems, and also increased the ability to rely on firearms to offer strong protection against cavalry and infantry attack. On the global scale, these changes possibly increased the effectiveness of Western armies more than their earlier adoption of gunpowder weaponry, an adoption shared across much of Eurasia. Relative capability, however, was not enhanced by a change shared by all armies, although it was against non-regular forces.

Within Europe, the key politico-military change in the century prior to the French Revolution was the rise of Russia. Indeed, Edward Gibbon captured its impact for the Eurasian system when, in his *Decline and Fall of the Roman Empire* (1776–88), he addressed the question of whether successful "barbarian" invasions could recur. Gibbon was most interested in this developmental characteristic, which challenged notions of cyclical change. Indeed, on the world scale, there had been a major change, one focused on the 1550s with Ivan IV's (Ivan the Terrible's) conquest of the khanates of Kazan and Astrakhan, taken on by successful expansion across Siberia to reach the Pacific in the 1630s, and, with greater difficulty, seen in Russia conquering the northern shores of the Black Sea, notably annexing Crimea in 1783.

This was yet another sphere in which the French Revolutionary and Napoleonic period saw continuity with what had gone before. Russia conclusively beat the Turks in the wars of 1768–74, 1787–92, and 1806–12, annexing Bessarabia (modern Moldova) as a result of the last. So also within Europe. If Alexander I, in the presence of allied rulers, reviewed a large Russian army at Chalons-sur-Marne in 1815 on the third anniversary of the battle of Borodino, Peter I (the Great) was in Copenhagen in 1716 with an army planning to invade Sweden, an invasion not in the event mounted. In 1716, 1735, 1748, 1798, and 1813, Russian troops moved into what is now Germany, in 1735 and 1748 helping to bring a war to a close, and in 1760 briefly occupied Berlin. A Russian fleet moved into the Mediterranean in 1769, another following in 1798.

In contrast to Russia's success in dominating Eastern Europe, a process that by 1815 had yielded Estonia, Latvia, Lithuania, Ukraine, Finland, and much of Poland, France was unable to dominate Western Europe. Looked at differently, Russia faced major challenges to its dominance, with both regional opposition and Western European intervention. Years of crisis for Russia included 1700, 1708–11, 1720–21, 1733–35, 1739–41, 1788–91, and

1806–12. In particular, combinations of crises, notably in the Baltic, Poland, and/or the Balkans, were of great difficulty, as was the prospect of Western European intervention, as in 1720–21, 1739, and 1790–91, a list that is not exhaustive. This situation serves as a reminder of the salience of strategy (or grand strategy), and notably the related issues of prioritization, and both for a power and for its opponents. These issues related to questions of skill in the management of coincident crises.

If Russia managed this successfully, there were also serious episodes, notably Charles XII of Sweden's invasion of Ukraine in 1708–9 and Napoleon's advance on Moscow in 1812. Both ultimately failed, and totally so, and, indeed, were less problematic for Russian stability than Polish intervention in the "Time of Troubles" in the early 1610s. The Poles controlled the Kremlin for much longer than Napoleon in 1812, but, for some reason, that does not influence perceptions of military history, whether of Eastern Europe or of Napoleon.

Compared to Russia's ability to dominate Eastern Europe, an ability that lasted until the German capture of Warsaw in 1915, and that resumed in 1944–89, France was unable to do the same in Western Europe other than in brief periods, notably 1806–12. That underlines the questionable character of focusing on France. Ultimately the "Western Question" that was posed from French advances, both into the lands of the Burgundian inheritance and into Italy, in the late fifteenth century, was ended in 1815, and with France failing. Moreover, the crisis was episodic between France's defeats in the War of the Spanish Succession in the 1700s and the victories of Revolutionary France from 1792.

The relative failure (compared to Russia) of France to dominate its region, let alone Europe as a whole, raises questions of explanation, about how best to assess military and political effectiveness, and concerning the causers of fears. In its region, China was far more effective than France. Answers can be offered, but how best to judge them is unclear. One argument rests on the more "multipolar" character of international relations in Western Europe, compared to Eastern Europe.

It is also appropriate to consider the ability of France's opponents to deploy effective forces as with Marlborough and Eugene in the War of the Spanish Succession, and with French defeats in Germany at Prussian (Rossbach, 1757) and British (Minden, 1759) hands during the Seven Years' War. In 1745–48, Marshal Saxe's victories in the Low Countries were not matched by the French in Italy or in the Rhineland. This ability to deploy effective forces against France remained the case into the late 1790s. Indeed, it was only in 1800–09 that the pattern was really broken. So also with Prussia, which, under Frederick the Great, far from necessarily succeeding found itself in serious difficulties from Russia and, eventually, Austria.

In trying to gauge the transition or relationship from the warfare of the *ancien régime* period to that of the Revolutionary period it is particularly valuable to focus on the last period of the *ancien régime*. To do so offers the great advantage that it is possible to avoid too static a picture of *ancien régime* warfare. That is especially valuable as it means that we thereby understand that it is not necessary to have a revolution in order to achieve substantive change, which indeed is a rhetorical rather than an analytical position. The "last period" of course is open to interpretation, in that the impression created may be very different depending on the years chosen. The decade ending with the outbreak of the French Revolutionary War in 1792 creates an impression centered on Russian effectiveness, while the two decades ending then give space to French success in the War of American Independence, which for France was 1778 to 1783. And so on. The variety of European warfare certainly emerges from a consideration of particular decades, and therefore the problematic nature of assuming a common trajectory.

Further uncertainty arises if trying to reduce the range of contemporary attitudes to one of the "Enlightenment view on war" or "the eighteenth-century view of war." For most contemporaries, there is no evidence of attitudes. A number of suggestions, however, can be made. First, war was normative and therefore it was likely not to have aroused attention as an unusual phenomenon requiring special attention. The role of religion was also such that attitudes toward particular conflicts as Christian or unchristian, and the views of religious bodies, were also probably significant. Given long familiarity with war, these were far more supportive of the process than is the case now.

To focus on prominent Enlightenment figures as if they defined the whole is therefore problematic. At the same time, it is useful; their views indicated a situation of flux well before the beginning of the French Revolutionary War. In particular, there were attempts to move the discussion of war from the contexts of Christian thought and royal *gloire*. These attempts spanned a range of writers from Voltaire to Adam Smith, and their preferences, and their views on rationality, were as contextual within the period as those attitudes they criticized. Nevertheless, the appeal to reason was one that looked forward to nineteenth-century views on war, albeit without the organic and biological metaphors and assumptions that were to become so important in its second half, and with scant sign that the causes of war changed. The idea of war in Europe as the guarantor of a balance of power was most clearly associated with William Robertson who, in his *History of the Reign of Charles V* (1769), wrote:

> That great secret in modern policy, the preservation of a proper distribution of power among all the members of the system into which the states are

formed. . . . From this era we can trace the progress of that intercourse between nations which had linked the powers of Europe so closely together; and can discern the operations of that provident policy, which, during peace, guards against remote and contingent dangers; which, in war, hath prevented rapid and destructive conquests.[19]

For Gibbon, who also praised the balance of power, military capability, in the shape of "cannon and fortifications," protected Europe from the risk of "barbarian" attack. Yet from the perspective of non-Western powers, there was little reason to be concerned about Western military capability on land. The major challenge the Chinese faced in the 1790s was that of rebellion, the White Lotus Rebellion, not foreign expansion, or even threat, and the Qianlong emperor was not receptive to British approaches in the Macartney Mission for cooperation. In the early 1780s, the British were under heavy pressure in southeast India from Mysore, while in 1779 Maratha forces had defeated the British in west India at Wadgaon. The Western states most challenging this situation were Britain and Russia, a position that did not change in the eighteenth or nineteenth centuries.

CIVIL WARS

The weakness, first of Safavid Persia and then of Mughal India, created opportunities that also led to conflict within. In India, "regional powers" emerged from within, gaining autonomy and de facto independence, as the Nizam of Hyderabad did in central India. In turn, this volatility encouraged conflict, as powers sought to define their position, and to understand and defend their interests in a situation made uncertain and threatening by fear and opportunity, emotions, attitudes, and practices shared by all. Thus, the Nizam fought the Marathas, who also benefited greatly from Mughal decline. From the 1760s, both came to compete with Haidar Ali of Mysore, who, like Nadir Shah in Persia and Taksin and Rama I in Siam, seized power through violence and then spent much of their reigns at war or preparing for war. The situation in late-Mughal India underlines the difficulties of employing a typology for warfare.

Even if these conflicts are excluded, it is clear nonetheless that civil warfare could be large-scale. In 1673, the Rebellion of the Three Feudatories began in China. Powerful generals who were provincial governors rebelled in response to the determination of the Kangxi Emperor to limit their power. This reflected the difficulties of managing patronage politics when the legitimacy of the new Manchu regime was poorly grounded. Order was not restored until 1681.

Possibly linked to the emphasis on strong states as the key players, there has been a tendency, as, misleadingly, with work on the twentieth century, to downplay civil warfare in the central narrative and dominant analysis of military history. Crucially, such warfare was not an important feature of Prussian military history, and notably not in the eighteenth century. The campaigning of Frederick the Great has long dominated attention for this period, and to a disproportionate extent. It is an aspect of American fascination with German war-making.

It is certainly mistaken to ignore civil warfare. Notably, the last quarter of the eighteenth century saw major insurrections and civil wars across Europe, including in Russia, Transylvania, Hungary, Belgium, the Netherlands, Ireland, and the Turkish Balkans, as well as in British North America, the Andean chain, Haiti, and China, especially the huge and very costly White Lotus rebellion of 1796–1805. Each insurrection was very different, and despite attempts to aggregate conflicts, as in the notion of a mid-seventeenth-century General Crisis or a late-eighteenth-century Atlantic Revolution, it is harder in these cases to draw common themes for the causes of war, the nature of conflict, or military capability, than it is when considering conflict between regular forces. Partly for this reason, but also owing to the relative lack of research, and to the prejudices of military establishments that saw (and see) internal policing and the suppression of rebellions as less prestigious tasks, this type of warfare has been relatively neglected in the literature.

CONCLUSIONS

Shaping this period into a general, let alone common, pattern of the causes of war is problematic. There is, as the chapter has indicated, a tendency to look at both sociocultural factors, which are focused on bellicosity, and at more specific political interpretations focused on a "rational" choice between options. The two approaches can both be treated as separate but, more helpfully, also brought together by stressing the crucial role of bellicosity in the process of choice. This can be seen for example in Chinese policy against the Zunghars, in the British move to conflict with Spain in 1739 and with France in 1755, and in Prussia's attacks on Austria in 1740, 1756, and 1778.

Again, this latter approach prefigures the analysis of the world wars in later chapters, suggesting that there has not been the development across time that it is otherwise overly emphasized. In this context, it is unwise to treat "the Westphalian system" as indicating a new practice or habit of mind. Indeed, peace settlements rarely reset attitudes toward war whatever the aspiration. The idea of drawing a line from Westphalia (1648), via Vienna (1815) and Versailles (1919), to the international architecture, notably the United Nations, that followed World War II is problematic, as well as Eurocentric to a degree that now appears curious and a subject for study, as well as wrong.

NOTES

1. Richard Tapper, ed., *The Conflict of Tribe and State in Iran and Afghanistan* (London: Routledge, 1983).

2. Gerald Jordan and Nicholas Rogers, "Admirals as Heroes: Patriotism and Liberty in Hanoverian England," *Journal of British Studies* 28 (1989): 201–22.

3. R. E. Glass, "The Image of the Sea Officer in English Literature, 1660–1710," *Albion* 26 (1994): 583–99; Alan McNairn, *Behold the Hero: General Wolfe and the Arts in the Eighteenth Century* (Liverpool: Liverpool University Press, 1997).

4. Philip V of Spain to Max Emmanuel, September 10, 1702, October 12, 1704, BN NAF 486 fols. 78, 89.

5. Paul Spalding, *Lafayette: Prisoner of State* (Columbia: University of South Carolina Press, 2010).

6. Frederic Baumgartner, *Declaring War in Early Modern Europe* (Basingstoke: Palgrave, 2011), 114.

7. Philip Mansel, "Monarchy, Uniform and the Rise of the Frac, 1760–1813," *Past and Present* 96 (1982): 103–32.

8. Marquis de Clermont d'Ambolise, French envoy, to Vergennes, French foreign minister, April 19, June 14, 1777, AN KK 1393.

9. V. G. Kiernan, *The Duel in European History: Honour and the Reign of Aristocracy* (Oxford: Oxford University Press, 1989).

10. Rory Muir and Charles Esdaile, "Strategic Planning in a Time of Small Government: The Wars against Revolutionary and Napoleonic France, 1793–1815," in *Wellington Studies*, ed. Chris Woolgar (Southampton: University of Southampton, 1996), 1:80.

11. Joanna Waley-Cohen, *The Culture of War in China: Empire and the Military under the Qing Dynasty* (London: Tauris, 2006).

12. Julius Ruff, *Violence in Early Modern Europe, 1500–1800* (Cambridge: Cambridge University Press, 2001).

13. H. Carl, "Restricted Violence? Military Occupation during the Eighteenth Century," in *Civilians and War in Europe, 1618–1815*, ed. E. Charters, E. Rosenhaft, and H. Smith (Liverpool: Liverpool University Press, 2012), 116–26.

14. Thomas Robinson, British envoy in Vienna, to William, Earl of Harrington, British Secretary of State, reporting views of Austrian minister Johann Christoph, Baron von Bartenstein, February 22, 1741, NA SP 80/144.

15. Richard Bonney, ed., *The Rise of the Fiscal State in Europe, c. 1200–1815* (Oxford: Oxford University Press, 1999).

16. Joanna Waley-Cohen, "Religion, War, and Empire-Building in Eighteenth-Century China," *International History Review* 20 (1998): 336–52.

17. Grant D. Jones, *The Conquest of the Last Maya Kingdom* (Stanford, CA: Stanford University Press, 1998), 300–2, 318–21, 329–35.

18. Keene to Charles Delafaye, Under Secretary, July 9, 1732, NA SP 94/112.

19. William Robertson, *History of the Reign of Charles V* (London, 1782 ed.), 1:134–35.

Chapter Five

Imperialism and Revolutions, 1790–1913

Within the West, revolutions were a key theme, leading, as they did, to the formation of nation-states, most clearly with Germany and Italy (or not, as with the American South), and to related wars, both international and civil. Yet, at the world scale, imperialism was most significant as a cause of transformative conflicts. This was very much the case in Africa, Asia, Australasia, and Oceania. Moreover, in the Americas, the key struggles prior to the mid-nineteenth century were wars of decolonization and the subsequent conflicts. A difference in tone is offered depending on what comes first for consideration, wars of revolution or wars of imperialism. Here, the approach is that of looking at wars of imperialism before considering struggles within systems, whether revolutionary or not. At the same time, classification, as ever, is highly significant, and leads to a discussion, at the outset, of the two important, but different, instances provided by, on the one hand, the French Revolutionary and Napoleonic wars and, on the other, the outbreak of the Anglo-American War of 1812.

FRENCH REVOLUTIONARY AND NAPOLEONIC WARS

The French Revolution that began in 1789 launched a warfare in 1792 that was at once revolutionary and imperialistic: revolutionary French forces, in seizing territory and establishing client regimes, such as the Batavian Republic in the Netherlands, were expanding the cause of revolution—in both theory and practice—and increasing French territorial hegemony and control.[1] Seizing power in 1799, Napoleon greatly continued the latter, but replaced the cause of revolution by his notion of modernization. His wars had

specific causes but also rested on his personality and ideology, and the bellicose nature of his regime. Napoleon controlled the French military system,
directed the war effort, and, throughout the wars of his reign, was in command of the leading French force in the field.

Partly as a consequence, Napoleon was unable to match his political goals
to the reality of a complex international system with which he needed to
compromise. For example, peace with Austria in 1801 and with Britain in
1802 left him dominant in Western Europe, but his determination to gain
advantages from the peace, his clear preparations for fresh conflict, and his
inability to pursue measures likely to encourage confidence led to a resumption of conflict with Britain in 1803. In turn, British efforts, notably the
payment of the subsidies made possible by the British system of public
finance and by British dominance of oceanic trade, led to the creation of the
Third Coalition against France.

Similarly, Tsar Alexander I of Russia, who had been forced to accept
French domination of most of Europe by the Treaty of Tilsit in 1807, was
increasingly concerned from 1810 about French strength, intentions, and
policies. Partly as a result, Franco-Russian relations deteriorated. As with his
earlier attacks on other powers,[2] Napoleon resolved to deal with a crisis he
himself had played a major role in creating by invading his opponent. Total
defeat by Russia in 1812 led, in 1813, to Prussia, Austria, and Sweden all
joining in against Napoleon.

Strategy is usually discussed in military terms, but that is simply to opera-
tionalize what is a set of political suppositions and drives. Indeed, Napoleon
created a situation in which the other European powers put aside their differences and united in a coalition pledged to fight him. This unity was largely a
product of the repeated bullying with which Napoleon had treated other
rulers in the 1800s and early 1810s and his willingness to tear up peace
agreements that he himself had dictated.

Conflict between governments could involve popular responses and thus
be the cause of a broader type of war. Rifts within Spain in the royal family
between Charles IV and his son Ferdinand, who enjoyed much support
among the higher nobility and the Church, rifts played out in public and
through publications, culminated in 1808 in the "Tumult of Aranjuez," a
coup at court that replaced the unpopular Charles, and his loathed first minister Godoy, with Ferdinand, who became Ferdinand VII. However, Napoleon
obliged both men to meet him at Bayonne where, under pressure, they had to
abdicate in favor of Napoleon. In a major extension of his imperial power, he
made his brother, Joseph Bonaparte, Joseph I of Spain. The response was
rebellion in the name of Ferdinand, notably in Madrid, and what is now
known as the War of Independence began.

This rebellion drew on a range of beliefs and tendencies, from liberalism
to conservatism, but there was a common reality of a rejection of French

sway. That was the cause of the war. Indeed, Spanish national identity was in large part to be defined in response to France. The *juntas* that took power in the provinces directed popular anger on the French, which provided a way to lessen political and social tension within Spain. This made it easier to recruit support, and notably so when not too much needed to be done. The insurgents against France were neither bandits questing for booty nor patriots simply fighting for crown, church, and country, although both elements played an important role. Instead, as in the region (former kingdom) of Navarre, they were also landowning peasants fighting to protect both their interests and a society that did so. The abolition of Navarre's privileges by the French under Napoleon, their efforts to reform society, and their rapacity all aroused great hostility, which was very much forced upon Spain by the French occupiers. At the same time, as so often, the geographical texture of politics, and therefore of rebellion, was more finely grained. In northern Navarre, there were more landowning peasants, but in the south of the province there were more day laborers and less harmonious social relations as well as more criticism of the Church. This ensured that opposition to Napoleon in Navarre was stronger in the north and therefore the rebellion more intense.[3]

The turmoil of the period in Spain, which included a partial breaking down of political and social structures, for example in the face of banditry and guerrilla activity, was more to the fore than a struggle simply against France. On the other hand, by reflecting and exacerbating a breakdown of governance, this turmoil ensured that the new French order was to be constrained by the need to enforce its presence. This was difficult, while control proved elusive.

More generally, the causes of war are in part a matter of choices between possible opponents. In 1793–95, Russia and, finally, Prussia, had preferred to concentrate on destroying Poland, rather than on fighting France. In contrast, in 1813–15, both powers focused on France. Moreover, in 1815, they settled serious differences over the future of the German kingdom of Saxony in order to do so. Napoleon's attempt to hold onto power ran up against not only the superior resources of his opponents but also the weight of recent history, which encouraged a prioritization on fighting him.

In March 1815, returning from exile where he had been sent after failure and abdication in 1814, Napoleon rapidly regained power in France. In letters to the Allied sovereigns of the Seventh Coalition, he pledged to observe existing treaties and affirmed peace with the rest of Europe, but his rhetoric within France toward the other powers was hostile and bellicose. Moreover, Napoleon sought a new league with the lesser powers, including Spain, Portugal, Switzerland, and the minor German and Italian states, a proposal that was a testimony to his lack of realism. So also was his confidence that the people elsewhere who had known his rule would reject war against France whatever their rulers thought. This diplomacy to peoples led Napoleon to order the publication of appeals to foreigners who had served in his forces to rejoin them.

Napoleon's march through France was not resisted by the forces of the alliance that had brought about his downfall in 1814, as none of these forces were still in France, but the powers determined on action. On February 26, Napoleon had escaped from Elba and, on March 20, he entered Paris. On March 11, the powers meeting at the Congress of Vienna knew of his arrival in France. Two days later they declared this an illegal act and offered help to Louis XVIII, the Bourbon ruler they had placed on the throne of France in 1814, and on the 25th they renewed their alliance in order to overthrow Napoleon, promising to support Louis.

British policy had to be defended from criticism in Parliament. On April 6, Robert, Viscount Castlereagh, the Foreign Secretary, told the House of Commons:

> It might be thought that an armed peace would be preferable to a state of war, but the danger ought fairly to be looked at: and knowing that good faith was opposite to the system of the party [Napoleon] to be treated with, knowing that the rule of his conduct was self-interest, regardless of every other considera- tion, whatever decision they came to must rest on the principle of power, and not on that of reliance on the man.

The context helped drive operations. In 1815, while the Allies assembled their forces, it seemed best for Napoleon to attack, benefiting from his posi- tion on interior lines. Attacking would enable him to disrupt the opposing coalition and would provide supplies from conquered territory. It would also enable Napoleon to strengthen his political position in Paris. On June 15, the war began when Napoleon invaded Belgium. Looked at differently, it had begun when he returned to France. On June 18, his forces were crushed at Waterloo.

THE WAR OF 1812

The War of 1812 was a conflict between two states, but, to its supporters in America, it was a conflict designed to bring the American Revolution to fruition both by conquering Canada and by wrecking Native American oppo- sition to American expansion, an opposition seen as resting on British sup- port from Canada. At the same time, another logic was provided by opposi- tion to Britain's naval superiority in the shape of the blockade of Napoleonic France. For that reason, the War of 1812 very much matched Napoleonic interests and has to be placed in this wider strategic context.

The causation of the war is a matter of historical debate. There is discussion of the respective roles of President Madison and of the War Hawks in Congress, headed by Henry Clay, Felix Grundy, and John Calhoun. Furthermore, it is difficult to disentangle and assess proximate causes from underlying attitudes. As far as the latter are concerned, a key element was that Thomas Jefferson, president from 1801 to 1809, and others overestimated American power after his success in acquiring Louisiana from France in 1803: the Louisiana Purchase took American claims to territory up to the Pacific Ocean, although, in practice, most of the territory was under Native American control. For sixty million louis, or $15 million, America gained over 800,000 square miles with no clear borders, including all or much of the future states of Montana, North and South Dakota, Minnesota, Wyoming, Colorado, Nebraska, Iowa, Kansas, Missouri, Oklahoma, Arkansas, and Louisiana. The United States also claimed the Oregon Territory (which included the modern states of Oregon and Washington as well as part of Canada) as part of the Louisiana Purchase.

The gain of Louisiana, including the crucial port of New Orleans, ensured that the Spanish stranglehold on the Gulf of Mexico was broken, challenging the Spanish position to east and west, in west Florida and Texas, while, as a result of the purchase, America now had a far longer frontier with Canada. Jefferson indeed tried to gain Florida in 1806. The new gains to the west were explored in 1805–7 by Meriwether Lewis and William Clark, who reached the Pacific coast at the Columbia River.

While Jefferson understood the potential of the West and was correct in his long-term appraisal that the United States would become a world power, he mistook America's marginal leverage in the bipolar dynamic between Britain and France for a situation in which all three were major powers, which was not to be the case until later in the nineteenth century. Madison, who had been Jefferson's secretary of state, followed this reasoning reflexively. This attitude ensured that American policy makers saw little reason to compromise in the disputes over neutral trade that played the key role in leading to war between Britain and the United States in 1812. Instead, they assumed that Britain would back down in the face of American anger and preparations for war, only to discover that they could not dictate the pace of events: Britain compromised, but inadequately, and too late, for the Americans.

Manipulated by Napoleon,[4] possibly as part of his attempt to unite all naval powers against Britain, Madison foolishly thought he had won concessions from France that justified his focusing American anger and the defense of national honor on Britain. To leading American politicians, Britain's maritime pretensions, which were more pressing than those of France, a weaker naval power, were a true despotism. In retrospect, John Threlkeld of Georgetown was to attribute the "wicked war" first "to the leaning of our government to that of France" and second to "the great desire of individuals to be general, colonel, captain, commodore, captain of frigates etc, privateers contractors etc."[5] In practice, French seizures of American shipping continued, while Napoleon's tyranny was more antipathetic than British policy to American values. Moreover, other powers, such as Sweden, moved away from France and toward Britain.

In an ahistorical moment, Richard Glover, a Canadian historian, compared the United States in 1812 to Mussolini's Italy, which in 1940 opportunistically joined Hitler's Germany against Britain when the latter was weak and vulnerable, as earlier it had been in the face of Napoleonic power in 1812. The American press certainly carried extensive news of France's naval buildup, which underlined Britain's vulnerability.[6] Indeed, once the war had broken out in 1812, Robert, 2nd Viscount Melville, the recently appointed First Lord of the Admiralty, expressed his concern about naval overstretch. Writing to Commodore Sir Home Popham, who was seeking more ships and sailors for his squadron off the northern coast of Spain, Melville noted:

> In addition to the American hostilities, which require to be met as they were likely to be of much longer duration than some people suppose, we have had since I wrote to you such demands for the Baltic and the Mediterranean as will completely drain us of all our disposable means, and still leave us deficient in many important points.[7]

It was also widely assumed that Napoleon would defeat Russia, not least as his invasion that year was supported by Austria and Prussia, each of which provided troops, as did lesser German powers such as Saxony and Bavaria.

A "visceral hatred of Great Britain" seems to have played a role for Jefferson and Madison.[8] Conversely, Spencer Perceval, the prime minister from 1809 until his assassination by a mad bankrupt on May 11, 1812, and Richard Marquess Wellesley, foreign secretary from 1810 to February 1812, who unsuccessfully tried, with the backing of the Prince Regent (the future George IV), to become prime minister in 1812, have been criticized for a "visceral anti-Americanism" that harmed relations.[9] Indeed, this hostility can be linked to the shift from an attempt to woo the United States to, instead, a tendency to adopt a firm attitude. Secretary of State Robert Smith complained in May 1810 of British "indifference" to relations and "inflexible determination" in persisting in its policies.[10]

The American misjudgment of Napoleon included an assumption that war between France and Britain would be bound to continue[11] and thus place Britain under pressure in any conflict with the United States. In fact, on April 17, 1812, Napoleon offered Britain peace, essentially based on the status quo. Suspecting an attempt to divide Britain from her allies, the British ministry, on April 23, sought clarification on the future of Spain, as they were unwilling to accept its continued rule by Napoleon's brother Joseph. Napoleon did not reply,[12] and that ended the approach.

Neither side probably was sincere in suggesting talks, but the possibility, however distant, of a negotiated end to the war in Europe, or at least one between Britain and France, was one that would have put a bellicose United States in a very difficult position. As a result, the international situation in 1812 was very different to the situation in 1778 when France had intervened in the War of American Independence. Britain had been greatly weakened as far as the Americans were concerned because France had gone to war with Britain while avoiding the European alternative offered by intervention in the War of the Bavarian Succession (1778–79), even though in this conflict France's ally, Austria, was at war with Prussia. In 1812, in contrast, the French turned east, albeit without ending their war with Britain, but this invasion of Russia was ultimately to lead to the downfall of Napoleon and end his war with Britain, and thus to weaken the United States.

Irrespective of French moves, Madison had departed from Jefferson's principles in foreign policy. Crucial to these were an attempt to maintain neutrality in great-power confrontation, which Jefferson presented as the way to avoid dangerous entanglements. This departure was to have a serious consequence for America. The domestic pressures on Madison, however, were serious and led Jeffersonians to fear for the survival of the republic. Unsuccessful as a tool of foreign policy, non-importation, the prohibition of British goods and ships, had also resulted in major economic strains, and this was increasing opposition to the government.[13] Indeed, having become much weaker, support for the opposition Federalists was rising. Meanwhile, Britain was not yielding over trade or impressments. Madison thus appeared to have the choice of backing down in order to assuage domestic pressures or of forcing Britain to back down. He underestimated the risks of the latter and failed to appreciate the prudence of the former. Backing down posed the more immediate risk, which is frequently a factor in a move to war.

In 1812, the British government itself did not want war with America.[14] For example, in 1811, Spain, then fighting as an ally of Britain against France, sought British help against American expansion into west Florida (the Florida Panhandle and over to the Mississippi, including coastal Alabama and Mississippi and part of Louisiana). The British government saw this expansion as unprovoked aggression against a close ally and instructed the envoy to protest, and to do the same if the Spanish colony of east Florida (modern Florida minus the Panhandle) was attacked. However, he was also ordered to avoid hostile and menacing language and informed that the government did not want to fight the United States.[15]

Madison, in his message to Congress on June 1, 1812, pressing for a declaration of war on Britain, emphasized maritime rights. He declared that Britain had been responsible for

a series of acts hostile to the United States as an independent and neutral
nation. . . . Thousands of American citizens, under the safeguard of public law
and of their national flag, have been torn from their country and from everything dear to them. . . . Our commerce has been plundered in every sea, the
great staples of our country have been cut off from their legitimate markets,
and a destructive blow aimed at our agricultural and maritime interests.

Thus, as far as Madison was concerned, war was already in progress. This
was a notion that was to dovetail with the idea that the new conflict was a
second "War for Independence," a view that made the war seem necessary to
some Americans, but that others rejected. Congress had already agreed to a
ninety-day embargo on exports in April. That June, Congress declared war,
with the Senate, which adopted the war bill on June 17, taking far longer to
persuade than the House of Representatives (adopted June 4) where the War
Hawks were powerful. Among the War Hawks, Henry Clay of Kentucky, the
Speaker of the House, was especially important. In advancing his views, he
also strengthened the role of the Speaker and used the speakership to advance
his cause.[16]

In part, the delay reflected the difficulties of justifying the war and the
bitterness of divisions. In the Senate, moreover, there was much support for
avoiding full-scale war, and, instead, limiting it to a naval conflict similar to
the "Quasi-War" with France in 1798–1800, namely privateering and action
against British shipping, in short a truly limited war that, in practice, was of
the form of an armed demonstration.[17] The "Quasi-War" is a reminder that
modern issues over the definition of war are not new. Initially, the Senate
voted for such a limitation, but it subsequently changed tack and backed the
House bill. Had the Senate not changed its mind, as indeed seemed possible,
then the House might have rejected its proposal, but that could simply have
led to an impasse that would have forced attention back to diplomacy. On
June 18, 1812, Madison signed the war bill into law.

That day, Augustus John Foster, who had been appointed British minister
plenipotentiary the previous year, was informed that the United States would
fight until Britain ended the impressment of sailors on American ships and
also granted the United States the generous understanding of neutral rights
offered to Russia in the Anglo-Russian Convention of 1801. This demand
ignored the extent to which Russia was a much more powerful state whose
support Britain had been seeking after the two powers had come close to war.

In 1812, the news of the declaration of war reached London on July 30. In
fact, the British government had already lifted the Orders in Council about
trade on June 23, in order to try to maintain peace and to encourage trade
with the United States, as well as in response both to French attempts to win
over the United States, and to economic problems within Britain. These had
been focused with a major campaign in the provinces, especially in Liver-

pool, the key port for American trade, for the repeal of the Orders in Council. This campaign reflected economic and political change in the shape of the rise in provincial economic interests and lobbying.[18] The news, however, only arrived in North America after fighting had broken out. Lifting the Orders in Council in relation to the United States would have come sooner had it not been for the assassination of Perceval and the resulting reconstruction of the government.

The British hope, once they knew that war had broken out, that the lifting of the Orders would lead to its speedy end, was misplaced. This mistake can be seen as a product of the focus of British Intelligence collection on Europe and the war with Napoleon. Admiral Sir John Borlase Warren, the commander in chief of a combined North American and West Indian station, was given instructions in August to negotiate with the Americans.[19] In practice, the British move was unacceptable to the United States as the British government insisted on the right to reinstate the Orders in Council when they chose, while there had also been no concession on impressments, which the American government insisted must be given up. On September 17, Castlereagh told Jonathan Russell, the American envoy, that impressments could not be ended. Meanwhile, on June 24–25, Napoleon's forces had crossed the River Niemen without resistance, launching his invasion of Russia. Over 600,000 French and allied troops were available for the invasion.

Congress had voted for war in part because of bellicose overconfidence, especially by the War Hawks, and in part thanks to patriotic anger with British policies, but very much as a party measure by the Democratic Republicans. Impressment of seamen from American ships was seen as a particular outrage, as it represented an infringement of the national sovereignty of American vessels and a denial of America's ability to naturalize foreigners, but impressment was not only a naturalization problem. Short of sailors in an age when warships required very large numbers to work the extensive sails and rigging and to operate their numerous cannon, British naval officers also impressed many native-born American seamen. This impressment was an extension of the forcible system used in Britain in order to raise sailors: the press gang. The maritime issues—Orders in Council and impressments—were crucial in the East, the region with most votes in Congress. However, for the New England ship owners, they were simply a cost of doing business that could be passed on to the Continental purchasers of American products. Impressment was emphasized by the bellicose Richard Rush, the comptroller of the treasury, in his Fourth of July speech at the Capitol that year. Rush linked this issue to the quest to affirm national independence.

Yet, to add to the difficulties in assessing causation, other ongoing tensions were also important. These included American suspicions of relations between Native Americans and the British in Canada, and the activities of British officials, officers, and traders on the frontier that, in large part, justified these suspicions. Until Jay's Treaty in 1794, the British had occupied bases such as Detroit, Mackinac, and Niagara, while there was a degree of connivance with the attempt by the Iroquois leader Joseph Brant to create a Western Confederacy to keep the Americans out of Ohio. This attempt failed in 1794,[20] but tensions continued thereafter, as did American suspicions. On the British part, there were still hopes that influence, commercial and political, could be developed in the interior of North America.[21] There was also a sense of obligation to the Native Americans as past allies, not least because Britain had let them down in its treaties with the United States in 1783 and 1794. The idea of the Noble Savage retained a hold on the British imagination.

The British, in turn, were blamed in the United States for problems with the Native Americans and with the fur trade. Both administrators and army officers in Canada indeed were prime movers in instigating the Native Americans to oppose American expansion. Many administrators were former officers who had fought the Americans, particularly John Graves Simcoe and Thomas Saumarez. However, the British were not responsible for the developing Nativist movement that centered on Tecumseh and, even more, his brother, Tenskwatawa, "the Prophet," who had launched a religious revival in 1805. This revival did not unite all Native Americans, and only a minority of the Shawnees followed Tecumseh and Tenskwatawa, but tensions between Native Americans and the United States rose. More generally, many Native Americans had little interest in accommodating themselves to American interests, and, instead, tended to see such a practice as weakness.[22]

Madison, in his message to Congress on November 5, 1811, urging it to prepare for war, made no reference to problems with Native Americans in his case for preparedness. Nevertheless, within the United States, the increasingly prominent (although only modestly represented in Congress) West, which then meant trans-Appalachia, especially Kentucky, Tennessee, and Ohio, was concerned about British aid to the Native Americans. This concern led to pressure for the seizure of Canada, which was seen as the base of, and the means for, this aid. Thus, expansionism arose in part from a defensive mentality. Prior to the War of 1812, the British government denied instigating Native American attacks on the Americans.[23] Insofar as such encouragement occurred, it was a case of policy being set by officials on the ground, especially military commanders, rather than by the foreign secretary and the diplomatic system. Indeed, the role of frontier disputes in relations with the United States provided the best example of the potential clash between formal diplomacy and other agencies of the British state. Again, the role of those on the spot in causing tension, if not conflict, was significant.

From the American perspective, the reality on the ground, however, was a degree of cooperation between Britain and Native Americans that was a threat. As a consequence, defeating the Native Americans was, in part, to the

Americans, a way of hitting Britain, whether or not the two powers were at war. In reality, Major-General Isaac Brock, from September 1811 the commander of the troops in Upper Canada, and the acting administrator of the colony, from the time he took command there, directed the Indian Department to exert its considerable influence over Native American frontier settlements in order to maintain the peace. Lieutenant-General Sir George Prevost, military commander in chief of British North America and governor-general, sent an extract of Brock's letter to the British minister in Washington in order to show how British authorities on the spot were acting to restrain the Native Americans. In May 1812, at the Mohawk village on the Grand River, Brock told Native American leaders that the British could not help against American encroachment while they were at peace.

American expansion, however, was a source of pressure to the Native Americans. Ohio became a state in 1803, the year of the Louisiana Purchase, which threw the status of the West and relations with Native Americans into greater prominence. The Greenville (1795) boundary rapidly became redundant as Native American lands in Ohio, Indiana, Illinois, and Michigan were purchased, particularly by the Treaties of Fort Wayne of 1809, which led to the acquisition of much of eastern and southern Indiana. In turn, this expansion led to a Native American reaction, which ensured that the fear of a renewed British–Native American alliance became a self-fulfilling prophecy.[24]

The Native American reaction helped cause fighting that was an immediate background to the War of 1812. Concerned about Shawnee opposition to the advance of American power, particularly opposition to land cessions by other tribes, William Henry Harrison (1773–1841), the governor of the Indiana Territory (and briefly president in 1841), advanced, in 1811, to seize their base at Prophet's Town near the confluence of the Wabash and Tippecanoe Rivers. On November 7, in what was subsequently known as the Battle of Tippecanoe, the Shawnees attacked first, but the early-morning assault by about 450 Natives under Tenskwatawa was fought off by about 910 Indiana militia. Tenskwatawa had misled the Native Americans by promising them immunity to American musket fire, a frequent theme in anti-Western movements that drew on religious inspiration. Harrison's robust defense exposed the claim and the outnumbered Native Americans withdrew, allowing Harrison to burn Prophet's Town and to claim a victory. Yet, it was soon apparent to Americans, Native Americans, and the British that the battle was not decisive. Indeed, Prophet's Town was repopulated and in June 1812 Harrison was to observe that the force there was as strong as that the previous summer.[25]

Tippecanoe indeed served to strengthen both the Native American determination to fight and Native American support for Britain, while also indicating the extent to which political issues were being militarized. In April

1812, Clay informed the House that "he had no doubts but that the late Indian War on the Wabash was executed by the British."[26] This was not the case, but it indicated the strength of a paranoid mood in the drive for war. Harrison himself had seen the treaties of Fort Wayne as simply the temporary prelude to fresh expansion, but ironically, the federal government was less than sympathetic, because Madison did not want a war with the Native Americans at the same time as he was pursuing the acquisition of west Florida from Spain, which he saw as more important. Jefferson, however, pressed for the conquest of Canada, informing John Adams in June 1812 that "the possession of that country secures our women and children for ever from the scalping knife, by removing those who excite them."[27]

During the War of 1812, the West was to feel that it was fighting an offensive-defensive war against British containment, a continuation of the struggles of the 1790s. To Westerners, this was not a case, as the critical, largely anti-war Federalists quipped, of fighting on land to protect maritime rights (a charge designed to expose the flaws in government policy), because, to them, the war was about much more. More widely, American expansionism had contributed to the crisis in Anglo-American relations, although it is possible that historians have overemphasized its role in bringing on the war, certainly insofar as Madison was concerned. This may indeed be an overemphasis that reflects historians' need to integrate maritime causes with internal or domestic ones.

Aside from the specific issues in dispute or arousing disquiet, there was a more general sense, particularly among the Jeffersonians, that the Revolution was unfinished because Britain remained powerful in North America, and that this power threatened American interests and public morality, as Britain was a corrosive, but seductive, model of un-American activity. This model moreover sapped American virility because it encouraged a trade that fostered a love of luxury, a love that corroded the necessary martial spirit.[28] The Revolution not only *seemed* incomplete; it *was* incomplete as long as Britain could be seen as a menace.[29] A sense of superior American morality, but also of the challenges it faced, was captured in 1812 by Thomas Sully's painting *The Capture of Major Andre*. Sully, who, as a reminder of Anglo-American links, had trained in London in 1809–10, depicted the three young militiamen who captured André in September 1780 refusing a bribe, and thus thwarting Benedict Arnold's plan to betray West Point. This refusal of the bribe, and the contrast with Arnold, was a symbol of the moral strength of ordinary American citizens—a strength that was the basis of the republic—but also of the need for vigilance against the British threat.

It was a highly divisive war, and one in which many Americans refused to help. The Federalists, heavily represented in New England, but not only there, were opposed to war with Britain, which they correctly saw as likely to be expensive and to harm trade. The Federalists also saw territorial expansion as intended to benefit their governing rivals. In addition, New England interests had only limited concern in the relations between Britain and the Native Americans that troubled the frontier regions. Commentators, indeed, had long detected regional rifts within the United States. Opposition in New England to the war looked ahead to limited New England support for the Seminole Wars in Florida, against the Native Americans there, and also for the Mexican-American War of 1846–48.

In 1812, Democratic Republicans outvoted the Federalists, but the divisions on declaring war—79 to 49 in the House of Representatives on June 14, and 19 to 13 in the Senate on June 17—reflected the depth of disquiet. These votes showed that it was difficult to create both a nation-state and a nationalism that worked. They also throw light on subsequent divisions in American history over going to war, underlining the extent to which they were part of a long tradition.

As the Constitution expressly conferred the power to declare war on Congress, and Congress alone could vote money to pay for the war and the military, the potentially unifying position of the presidency was heavily qualified thereby. Moreover, it was too early for the potential wartime powers of the presidency to be explored fully. They only became clearer with the Civil War fifty years later.

The prime American strategic goal was the conquest of Canada, an issue that symbolized the extent to which the American Revolution seemed unfinished abroad and, therefore, in some eyes unrealized at home. It is unclear, however, whether, once conquered, Canada would have been retained by the Madison government or merely used as a pawn in negotiations with Britain over impressments and trade. Its use as a negotiating pawn has been stressed in accounts that emphasize the defensive character of American policy; but, whether defensive or offensive, these goals required the seizure of Canada. So also, in the eyes of the many Westerners, did any settlement of the Native American issue. Clay was to write in December 1813, "When the war was commenced Canada was not the end but the means; the object of the war being the redress of injuries, and Canada being the instrument by which that redress was to be obtained." However, he also observed that, if Canada could be conquered and retained, it should be held.[30] That was very much a theme of the Westerners.

The complex nature of sectional interests was a factor in American politics in 1812. Far from being focused solely on Canada and Native Americans, the British Orders in Council were blamed by the Western and Southern War Hawks for agricultural problems in their regions. Thus, alongside the issue of national honor, the need to force Britain to change maritime policy was also seen there as a key sectional interest. Maritime issues, in short, were not a diversion from Western and Southern concerns about Native Americans.[31]

WARS OF IMPERIALISM

The period from 1816 until 1913 is sometimes presented as relatively peaceful in terms of the numbers of conflicts per year, the degree to which individual states were involved in war, and the length of time when major powers did not fight each other—all in comparison both with the preceding century and the subsequent half-century. This is misleading, even as a description of Europe, because it ignores both the high level of civil warfare and military intervention in the domestic affairs of other states, especially in Europe from 1816 until 1849, and the degree to which many European powers were engaged in colonial warfare.

Much of the warfare of this period indeed arose from the worldwide expansion of European power, but it would be misleading to suggest that the sole expanding powers were European or, indeed, that most of the warfare of the period involved Europeans. Prior to the 1840s, the European impact on much of Asia was limited, and this was also the case of much of Africa before the 1880s. In fact, the list of expanding powers in the nineteenth century included, in Africa, Egypt, Lunda, Abyssinia, Sudan under the Mahdi, and the Zulus, as well as a series of personal empires in West Africa, notably that of Samori Touré (c. 1830–1900), the "Napoleon of the Sudan," a term employed by Europeans in order to place him and to provide glory from his failure. The context and character of such expansion varied greatly. For example, the jihad launched by Usman dan Fodio in West Africa in 1804 was directed against other Africans, especially the—in his eyes—insufficiently rigorous Islam of Hausaland. The result was the creation of the Sokoto caliphate. In contrast, for a short while, Mehmet Ali of Egypt (r. 1811–49) must have been one of the most successful conquerors of the century. He was motivated by a drive to create a powerful "modern" state and a new dynasty. Egyptian power was extended into Arabia, Sudan, Israel, and Syria. In turn, Egypt was defeated both by the jihadist Mahdi and by Britain. Touré, a Guinean Muslim cleric, created the Wassoulou empire in West Africa in 1878, resisting France from 1882 until his capture by the French in 1898. In Asia, China was still able to intervene effectively in Nepal in 1792, and the list of expanding states there would include Burma, Punjab under the Sikhs, Siam (Thailand), and Japan.

In the nineteenth century, the ambition and capability of European states for imperial expansion in Africa and sub-Siberian Asia increased. This shift was emulated by states that Europeanized at least their armed forces, notably Egypt and Japan. Long-standing imperial powers greatly increased the intensity of their territorial ambitions. In Britain, Parliament might pass the India Act of 1784, which declared that "schemes of conquest and extension of dominion in India are measures repugnant to the wish, the honour, and policy of this nation." Even so, in the ten years prior to Britain's entry into the

French Revolutionary War in 1793, Britain did make gains: in Australia, Sierra Leone, southern India, and at Penang in modern Malaysia. In the case of European powers, colonial disputes between them might well lead to war. Thus, in 1790 a confrontation between Britain and Spain over the Pacific coast of North America, the Nootka Sound Crisis, nearly led to war with Spain and its ally France. However, the mechanisms for negotiation and compromise existed, because relations with other European powers operated in accordance with familiar diplomatic conventions.

In contrast, even though extra-European territorial gains at the expense of non-European powers could be limited, the mechanisms for establishing a compromise settlement were less ready, and policy was not in the hands of diplomats seeking a compromise. Thus, war with Spain was avoided, but not, in 1790, war with Mysore, the Third Anglo-Mysore War. The outbreak of the "Opium War" between Britain and China in 1840 reflected the difficulty of reaching compromise. As a reminder of the problems of determining what exactly is war, it is also dated to 1839.

In the case of imperialism, there is, yet again, the interplay today of an understanding of a bellicist political culture (a term itself with a variety of applications) with a consideration of why particular wars broke out. However, as with Christian-Ottoman relations in the sixteenth century, this can be a misleading approach, for, in some regions, a sense of conflict, or at least dispute, was continuous. Therefore, the notion of "war" as something distinct was, and is, problematic. This was in part true of American and Brazilian expansion into "their interiors," and of Russian expansion into both the Caucasus and Central Asia. In these cases, and more generally, aggressive local officials, especially military commanders, land speculators, and traders, could greatly exacerbate relations. A contrast between national policy making and more aggressive military initiatives is readily apparent in the case of the United States in the West.

At the same time, and on a long-standing pattern, there were domestic and international factors that encouraged expansion and conflict at particular junctures. British expansionism in India exemplifies this process. For example, a forward drive in policy in the late 1790s arose in large part from the determination of a number of officials, especially Governor-General Richard Wellesley, but it was defended as necessary to prevent French intervention in India: Britain and France were at war. Against this background, Indian territories were seen as dependents whose foreign and domestic policies were to be controllable by British officials. This was unacceptable to Indian rulers and led to a series of conflicts beginning with the Fourth Anglo-Mysore War in 1798–99 when Tipu Sultan refused to accept demands that would have made Mysore a British protectorate. Conquests also increased tax revenues, helping to relieve an increasingly hard-pressed East India Company and to finance military activity.

Conquest as forward-protection, a form of protective war, is a major theme in this and other periods. In addition, imperial wars in the late nineteenth century have been seen both as the products of a desire for economic control over the world and as a result of competitive deadlock within Europe, particularly after 1871. In a parallel point, stasis between the United States and the Soviet Union in Europe during the Cold War encouraged aggressive activities elsewhere. For fear of a nuclear war with the Soviet Union, the United States could not practice "roll-back" in Eastern Europe, notably in Hungary in 1956, but could seek to pursue it in Korea and Vietnam.

During the nineteenth century, the major European powers certainly competed in part by expanding their influence and power in non-European parts of the globe. This was a sphere where their rivalries could be pursued and prestige gained with a measure of safety, certainly compared to conflict in Europe, and without too substantial a deployment of resources. The competitive element had been important from early on. Thus, France attacked Algiers in 1830 and in 1834 also decided to conquer the Algerian littoral, partly in order to challenge British dominance in the Mediterranean. Alleged British ambitions were also used to justify French expansion in West Africa in the 1880s and 1890s, and German moves encouraged French expansion in Morocco in the 1900s.

In the case of Britain and Russia, the crushing Russian naval victory off Sinope in 1853 provoked concerns about the weakness of the Turkish empire and about possible Russian domination in the Black Sea and the Balkans. Employing the geopolitical ideas of the day, British commentators saw this as a threat to the British overland route to India, although they considerably exaggerated this threat. In practice, Tsar Nicholas I of Russia paid more attention to continuing the earlier Russian expansion at the expense of the Turks. Napoleon III of France, however, saw both crisis and opportunity in the situation. By intervening, he could achieve prestige to strengthen his domestic position (he had recently made himself emperor) and to enhance France's diplomatic situation. The British government greatly distrusted Napoleon III and his intentions; nevertheless, distrusting Russia even more, it joined with France in seeking to limit Russian expansion in the Crimean War of 1854–56.[32]

Colonial expansion by the leading powers encouraged weaker European empires to be aggressive, with the Dutch conquering Aceh in Sumatra, Portugal resuming expansion in Angola and Mozambique after a gap of three centuries, and Spain seeking to expand in Morocco. This expansionism seemed the best way for them to prevent despoliation by the leading powers: to conquer was to be. Under General Leopoldo O'Donnell, who governed in 1856, 1858–63, and 1864–66, Spain, despite its large state debt, was involved from 1859 in a series of imperial episodes that were intended to ensure public support and that reflected the wish to act as a great power, as well as a sense that Spain should still have a role in Latin America. O'Donnell was an interventionist with a strong penchant for dramatic gestures on the international stage. Spain participated in French-led interventions in Vietnam and Mexico and, on its own, in response to French expansionism in Algeria, in a campaign in Morocco in 1859–60 that led to the capture of Tetuán and Ifni in 1860. In the former empire, there was resumption of control of Santo Domingo (now the Dominican Republic) in 1861. However, American pressure in 1866 helped lead Spain to end naval hostilities with Peru and Chile. The pressure was short of conflict, as also with that on Napoleon III to withdraw his forces from Mexico, which he did. In 1865, Spain had abandoned its attempt to control Santo Domingo, which, as a result of local opposition, had turned out to be a fruitless commitment.

Other rising states sought to conquer in order to pursue their "Place in the Sun" and catch up. This motivated both Germany and Italy, especially after the Congress of Berlin in 1884–85 encouraged a fear of a closing door on opportunities for imperial expansion. This fear led them into war, Italy against Ethiopia and later Turkey, and Germany into conquest and then harsh counterinsurgency struggles in Southwest and East Africa.

By the late nineteenth century, expansion was fostered because there were far more points of European entry outside Europe than had been the case earlier. Moreover, the earlier emphasis on mutual benefit, as well as rivalry, in relations between major empires and their "barbarian" neighbors was thrust aside by the Europeans, notably in the 1880s and 1890s. Yet, it was equally true that European imperialism owed much to divisions among the non-Europeans, as in India, and indeed to their military support. Both could be crucial, for example in West Africa and South Asia, but the decision-making process was very much kept in European hands.

In general, war between European powers over transoceanic expansion was avoided, especially after 1815. Prior to that, for example, Spain and Russia had avoided conflict in the Pacific. Subsequently, the European powers cooperated in the carve-up of Africa and other parts of the world, especially in the Pacific. Sustained disputes were less common than treaties that agreed boundaries between colonies. Despite several war-panics, in which both sides prepared for conflict, there was no war between Britain and Russia after the Crimean War of 1854–56. Thus, in 1885, the two powers prepared to fight, after the Russians, as part of their drive to subjugate Central Asia, seized Merv and advanced south toward Afghanistan, trying to extend their hegemony over the tribes of the region, in defiance of what the British regarded as an agreement between the two states. The British India army prepared for action, and fighting took place between the Russians and an Afghan force that had advanced north with British encouragement and advisers. Nevertheless, a territorial compromise was eventually negotiated and the

British India army did not fight the Russians. The buffer zone remained in place.

However, there were conflicts between European powers and states formed by people of European descent—the Anglo-Boer wars of 1881 and 1899–1902, the Spanish-American War of 1898, and French intervention in Mexico in 1861–66—and between these peoples and Europeanizing states: the Russo-Japanese War of 1904–5. These conflicts stemmed, in part, from the prestige of imperial states that felt themselves unable to make compromises or accept losses. Thus, in 1898, the Spanish government and army could not face the prospect of abandoning Cuba, although they were aware that the United States was a formidable foe.

The Spanish-American war also needs to be located in struggles between Spain and indigenous resistance. In Cuba, there were two major insurrections. The first, in 1868, was linked to the crisis caused by the overthrow of Queen Isabel II in Spain: the cause was both specific to the colony and more general to the Spanish world, a situation that repeated that earlier in the century in Spain's mainland American colonies. As also there, civil warfare was involved. Spain benefited greatly from ethnic, geographical, and social divisions among the Cubans, and from the willingness of many, in part as a result, to support Spain. In particular, from 1870, Cuban whites increasingly rejected what they now saw as a black-run revolution focused on opposition to slavery. Once civil war in Spain, the Third Carlist War, ended in 1876, 25,000 troops were sent to Cuba, and control was restored in 1878.

The insurgency in Cuba resumed in the mid-1890s. In turn, successful American intervention in 1898 led to the end of Spanish rule. However, in the Philippines, American forces, having defeated Spain, found themselves in conflict with an indigenous independence movement, thus creating a new war, or rather new combatants for an old one.[33]

In the case of the Anglo-Boer Wars, British leaders found it difficult to accept Boer views and were willing to risk war in order to achieve a transfer of some power in southern Africa. The Anglo-Boer War of 1899–1902, the major struggle, is often seen as a classic example of "capitalist-driven" empire building. However, many capitalists with interests in the region concerned themselves with politics and war only when aggressive business methods did not meet all their needs. In contrast, Alfred Milner, the aggressive governor of Cape Colony from 1897 to 1901, who stood in a long line of what Disraeli called "prancing proconsuls," was essentially driven by political considerations and his own ambition. Ministers in London thought the Boers were bluffing and would not put up much of a fight if war followed, while the failure of the British to send significant reinforcements persuaded the Boers to think it was the British who were bluffing. The Boer republics (Orange Free State and Transvaal) declared war after Britain had isolated them internationally and had done everything possible to provoke them.

The British ministers were greatly influenced by the fear that if, given the gold and diamond discoveries, the Boers became the most powerful force in southern Africa, it might not be long before they were working with Britain's imperial rivals, especially the newly established Germans in southwest Africa, and threatening Britain's strategic interests at the Cape, the key port on the oceanic routes to India and Australasia. The prime minister, Robert, 3rd Marquess of Salisbury, no mean imperialist, remarked that Britain had to be supreme. Although the Boer War might never have happened had it not been for the gold and diamonds upsetting the economic balance of power, it is necessary to be cautious before ascribing too much to the capitalists: those in business were less important than the capitalists in government, and the latter were concerned about power rather than business.[34]

Competing Russian and Japanese interests in the Far East interacted with domestic pressures, including the view in certain Russian governmental circles that victory would enhance the internal strength of the government, and a foolish unwillingness to accept Japanese strength, interests, and determination, a view that prefigured Japanese assumptions about the United States in 1941. Engaged in a complicated struggle for power and influence in St. Petersburg, the Russian leaders were too divided among themselves, and too intent on their own power struggles, to have a coherent Far Eastern policy. The finance minister was very opposed to war, as finance ministers are apt to be. However, many of the military had served in the Far East; they were linked to commercial adventurers with interests in the region—the Russian counterparts of Cecil Rhodes; and they were close to the tsar, Nicholas II. The government did not seem to have been looking for war, but it failed to see that serious dialogue with Japan was necessary if war were to be avoided. The tsar and his advisers did not think the "yellow devils" would dare to fight. Russian behavior and arrogance did much to create ultimate unity in Tokyo in 1904 behind war. There was also timing and the related matter of strategic potential.[35]

Governmental attitudes were crucial in decisions for war, but political and popular attitudes and pressures were also important. Nineteenth-century imperialism drew on a variety of attitudes, including Social Darwinism and theories of racial superiority, none of which encouraged a sympathetic treatment of non-European views. The risings in the 1900s in Namibia and Tanzania against German control led to debate in Germany over imperialism. In the *Reichstag*, however, Matthias Erzberger was jeered when he announced that Africans had souls. Imperialist sentiment, indeed, led the critical Social Democratic Party to lose votes in the 1907 election—the so-called Hottentot election, just as in Britain in 1900, in the "Khaki election" during the Boer War, the more bellicose Conservatives had won votes at the expense of the divided Liberals, a result that kept the Conservatives in power for longer than otherwise might have been the case.

Confidence in imperial mission and military strength encouraged a resort to force. Although humanitarian considerations played an important role in Western attitudes and policies, as in Christian missions and in moves against the slave trade, these considerations arose from the potent interaction of a conviction of superiority and a strong sense of mission. Thus, in Algeria, invaded by France in 1830 largely in order to win prestige for the Bourbon monarchy, Christianity, civil law and education, and a secular state were all introduced. A short-term expedition became a long-term commitment, one that in French terms was that of extending civilization.

The purposes and parameters of European overseas diplomacy were thus different to those affecting international relations within Europe. War was not a product of the breakdown of diplomacy, as much as an activity that made as much, if not better, sense of the nature of international relations. In Algeria, the French progressively pushed their control south toward, and into, the Sahara. The Russians, who also had a view of themselves as extending Christian civilization, justified their expansion into Central Asia on the grounds of the need to establish a stable border. Viscount Curzon discerned a physical inevitability in the Russian advance: "in the absence of any physical obstacle and in the presence of any enemy whose rule of life was depredation, and who understood no diplomatic logic but defeat, Russia was as much compelled to go forward as the earth is to go round the sun."[36] Expropriation and continual pressure characterized the treatment of the indigenous population in the United States, Australasia, and Latin America, as with the "Conquest of the Desert" campaigns in Argentina in the 1870s. Treaties with non-Western peoples and polities were imposed, ignored, or arbitrarily abrogated. So also with the Japanese treatment of China: contempt, hostility, and racialism all played a role.

In the treatment of non-Western powers, there was no sense of mutuality in international relations. This was clearly seen in 1798 when Napoleon's invasion of Egypt, still part of the Ottoman Empire, almost casually led to war with the Turks. The French had invaded Egypt in order to be better able to challenge the British position in India. They had assumed that the Turks could be intimidated or bribed into accepting French action, which followed a series of provocative acts. These assumptions were coupled with a contempt for Turkey as a military force. Napoleon's sense of grandiloquence and his belief that the Orient was there to serve his views emerged from his recollection:

> In Egypt, I found myself freed from the obstacles of an irksome civilisation. I was full of dreams. . . . I saw myself founding a religion, marching into Asia, riding an elephant, a turban on my head and in my hand the new Koran that I would have composed to suit my needs. In my undertakings I would have combined the experiences of the two worlds, exploiting for my own profit the

theatre of all history, attacking the power of England in India and, by means of that conquest, renewing contact with the old Europe.[37]

Reality was to be otherwise. The Turks resisted and, in alliance with Britain and Russia, did so successfully. The French cultural supposition of superiority and arrogance of power had led to a lack of sensitivity that caused the war. Although it was Selim III who declared war on France, there was no viable alternative response to the French invasion of Egypt.

Europeans were apt to see their activity in terms of a struggle between rising (themselves) and declining states, as when Italy invaded Turkish-ruled Libya in 1911, in the event with only partial success. At the same time, there was also conflict between expanding polities. The First Anglo-Burmese War of 1824–26 began with a declaration of war by the East India Company because the aggressive, expansionist kingdom of Burma, keen to consolidate its frontiers and end disorder in neighboring principalities, clashed with the company's fears and its defensive determination to protect British protectorates in northeast India. Burma and the company had a common frontier as a result of the Burmese conquest of Arakan in 1784. The unsettled nature of the frontier, and the disruptive role of Arakenese refugees, gave rise to serious disputes and created distrust. When, in the 1850s, the French attacked the Tukolor empire of al-Haji Umar in West Africa, it was expanding to the east at the expense of the states of Kaarta, Macina, and Segu.

Even if, at a particular moment, the European power did not want a full-scale conflict, the general tenor of its policy was such that it could be seen as aggressive and a threat. When the situation appeared propitious, a violent response could be launched, as by the Turks in 1787. The suspicions of the British held by Sher Ali of the Afghans led to the Second Anglo-Afghan War of 1878–80. That such a movement was often one of hesitation, even conciliation, on the part of the imperial power apparently confuses the issue of responsibility, but the inherent cause in such cases was the general aggressive attitude and policies of such powers. This was a parallel to the question of how far Napoleon and Hitler were responsible for the wars in which they were involved.

Furthermore, Western advances led to religious-cultural disquiet and disorientation that produced movements of religious reaction, as among the North American Natives; the Muslims of Algeria, Libya, Sudan, the Caucasus, Sumatra, and the Northwest Frontier in the 1890s; and in sub-Saharan Africa, as in Matabeleland in 1896 and in Tanzania in 1905–7.

The greater control of states over their peoples and the extension of the world of officialdom into border zones ensured that government agents and, in most cases, the agents of central government played a greater role in frontier regions and border warfare than in earlier centuries. At the same time, this increased the responsiveness of central government to the situation on distant frontiers. Moreover, the men on the spot continued to be significant. In West Africa, the *officiers soudanais* of the Marine Corps dramatically increased French territory in the 1880s and 1890s, despite a lack of governmental support. Russian commanders in Central Asia were similarly responsible for aggressive action. In both cases, the incoherence of the policy-making machine was important. In the case of Britain, Sir Bartle Frere, governor of Cape Province and commander in chief for South Africa, adopted an aggressive stance toward the Zulus in 1878, launching an offensive the following year against the wishes of the colonial secretary. Similarly, Brigadier-General Frederick Lugard, High Commissioner for the Protectorate of Northern Nigeria, was pressed by the British government to keep the peace with the Sokoto Caliphate. Nevertheless, determined on war, he got it in 1901, the pretext being the murder of a British officer and the failure to hand over the murderer. Other justifications included the continuation of slave raiding.

The advance of Western power brought Western concepts of sovereignty, suzerainty, frontiers, and conflict into dispute with those of other societies. That was not the key problem within the West, however, apart from the major difference posed by the French Revolutionaries.

THE MEXICAN-AMERICAN WAR

Conflicts between independent Western powers in the New World could provide an interesting overlap with wars of imperialism. The Mexican-American War of 1846–48 certainly saw a form of racialism on the American part. Mexico was treated as backward and Mexicans were despised, in part because they were Catholic. American expansionism represented a continuance of earlier expansionism at the expense of Spain. At the same time, there was a focus on the American side on central government action in 1846 whereas some of the earlier activity against Spain, and then Mexico, had reflected local initiatives, notably the practice of filibustering. Opportunity in the shape of an assessment of relative possibilities was also important.

The 1844 presidential campaign had committed the victorious President James Polk to the goals of advancing to the Pacific, to the "re-annexation" of Texas, as well as the "re-occupation" of Oregon. The unresolved nature of disputes over Texas provided the opportunity for the American government to move to force with Mexico, the defeated inheritor of Spain's imperial position in Texas, once diplomacy had failed. The British sought to mediate a settlement between Mexico and Texas in order to help maintain Texan independence from America, which was regarded as a way to limit American expansion and to provide an independent source of cotton imports from outside the American tariff system. A treaty of amity and commerce with Texas was ratified by Britain in June 1842. However, the British government did not wish to offend the Americans by formally opposing annexation.[38] In 1845, Charles Bankhead, the British envoy in Mexico, helped persuade Mexico to recognize Texas, a step it had refused to take for nine years. Texas, however, did not fulfill British hopes; it preferred union with America that year, despite Elliot's effort (backed by his French counterpart) to persuade it to preserve its independence on the basis of Mexican recognition. In contrast, the American government was willing to provide the offers of assistance against Mexican invasion that the Texas government sought, and American troops were ordered into Texas in June accordingly. The decision for union was the key step, with a Texas convention voting on July 4 to accept the invitation by Congress that February to enter the Union. The vote led to strong pressure in Mexico for war, as this union was seen as a threat to Mexican national honor and as sealing the Texas Revolution against Mexico in 1835–36. Texas was admitted to the union on December 29, 1845.

In addition, there was continuing disagreement over the Texan frontier, with Mexico refusing to accept the Rio Grande, the border extorted from the captured Mexican commander Antonio López de Santa Anna after his defeat at San Jacinto, as part of the Treaty of Velasco in 1836. Particular controversy attached to the Nueces Strip, the territory between the Rio Grande and the Nueces River to its north, which was seen by the Mexicans as in the Mexican state of Coahuilla, while, farther north, Santa Fe and the land to the east of the Rio Grande were part of the province of New Mexico. Despite Polkite claims, the Strip was not part of Texas, as Abraham Lincoln understood, when he dared Polk to prove the claim. American Whigs generally supported Lincoln's position.

Believing that Mexico would back down, and keen to show America as a resolute supporter of Texan interests, Polk proved willing to press for the Rio Grande as the border. He authorized the deployment of troops south of the Nueces River to provide security for Texas, and tried to purchase the disputed territory, New Mexico, and California from Mexico for $35 million; but there was to be no second Louisiana Purchase. The Mexican government rejected the offer, and also refused to receive John Slidell, the American envoy. Polk tried to increase pressure by sending an army under Zachary Taylor to the Nueces, but the Mexicans responded by sending their own army to the Rio Grande. For a while, both sides stayed out of the Nueces Strip, but the failure of negotiations led Polk to order Taylor to the Rio Grande, which was an invasion as far as Mexico was concerned. He moved south in March 1846.

The result of the subsequent war might seem foreordained, in part as a result of the contrast in political culture and resources between the two sides, as well as reading back from the eventual result; but this is a problematic approach. Mexicans hoped that the Americans would be weakened by internal political divisions, slave insurrection, Native American opposition, and British hostility arising from the Oregon issue, as well as by the deficiencies of a small army and low-quality volunteers. They were also sure of the strength of their own military. Waddy Thompson, the American envoy, however, claimed in the early 1840s:

> I do not think that any commander could perform a tactical revolution with five thousand Mexican troops. I do not believe that such a one—a manoeuvre in the face of any enemy—ever was attempted in any Mexican battle; they have all been more melees or mob fights, and generally terminated by a charge of cavalry, which is, therefore, the favourite corps with all Mexican officers. [39]

In early 1846, there was a widespread expectation in America and among its army that a weak Mexico would back down, an expectation in which Polk was encouraged by Colonel Alejandro Atocha, a representative of the exiled General Santa Anna, Mexico's former leader. However, in late December 1845, General Mariano Paredes and his army overthrew the government of José Joaquin de Herrera, in part because the government was discredited by reports that it was yielding to the Americans by proposing to sell the disputed territory, as well as New Mexico and California. Richard Pakenham, the British envoy in the United States, reported in May 1846 that

> the Americans greatly underrate the difficulty and expense of a war with Mexico. Unless the Mexican character has undergone a great change since I left that country, I think the Americans will meet if not with a gallant resistance at least with a sullen and a dogged resolution to protract the struggle to the utmost, were it only for the sake of the expense and embarrassment which such a contest must occasion to this country. [40]

He was also unconvinced about the strength of the American military:

> Numerous bodies of volunteers have been forwarded to Matamoros from New Orleans and other places, but apparently in a hasty and confused manner, and, as far as I can discover, without any attempt at discipline or organization. It seems difficult to imagine how a campaign undertaken under such circumstances in an enemy's country can be successful, however low an estimate may be formed of the resistance to be encountered. [41]

Conflict started on April 25, 1846, when an American patrol on the north bank of the Rio Grande was ambushed by Mexican cavalry seeking to challenge the American presence, but both governments had already determined on war. The Mexican government had decided that Taylor's advance meant that a state of "defensive" war existed, and, on March 8, Paredes told Bankhead that America was an "indefatigable and powerful neighbour" and that he did not want Mexico to become her prey. [42]

Claiming the need to repel invasion, Polk falsely informed Congress that he had tried to avoid hostilities, and secured a declaration of war. This course of action attracted criticism at the time and has been commented on more recently in the context of misleading presidential statements helping lead to war. Following approval of his war bill by the House of Representatives on May 11, by a vote of 174 to 14, and by the Senate the following day, by a vote of 40 to 2, Polk signed the measure on May 13.

The expansionism of 1840s' America was also directed against the British position in North America, in the shape of the Oregon Question. There were calls for occupying the entire territory, which included what is now British Columbia. Polk indeed campaigned successfully for president on that basis. However, in this case, bellicosity ran into the roadblock of prudence, a process helped by Britain's willingness to be accommodating and by the better alternatives offered by expansion at the expense of Mexico. Even so, the latter war indicated the potential divisiveness of the turn to conflict. As with the War of 1812, there was strong opposition with a clear regional component: the South did not wish to see an extension of Free Territories, and the Northeast did not wish to see an extension of territories open to slavery. Thus, an important aspect of the causes of war was its politics in the shape of the creation of a constituency willing to support the turn to war. Looked at differently, the turn reflected the constituency. In 1846, the Oregon Question was settled by a territorial compromise, which left Britain with the future British Columbia and the United States with the future Oregon and Washington.

WARS OF REVOLUTION: THE CASE OF THE LATIN AMERICAN WARS OF INDEPENDENCE

There was no clear separation between wars of imperialism and wars of revolution. The "Mutiny" or "War of Independence" in India against British rule in 1857–59 fully shows that. At the same time, there were wars of revolution within, as opposed to across, cultural areas. These wars could involve simply conflict in the state in question, most prominently with the American Civil War of 1861–65 and the Taiping Revolution in China of 1851–61. However, they could also involve broader wars due to foreign intervention, as with the outbreak of the French Revolutionary Wars in 1792. Moreover, a revolution once successful could lead to a state that, as a sequel, sought to expand or, as it viewed it, complete the revolution, as with the American attack on Britain in the War of 1812 discussed earlier in the chapter.

Wars of revolution between 1775 and 1825, in the United States, Haiti, and Latin America led to the collapse of European control over most of the New World. They reflected the role of chance events, notably Napoleon's takeover of Spain in 1808, which greatly disrupted the Spanish American world. In Spanish America, in 1809 and 1810, local *juntas* seized power in the name of Ferdinand VII, underlining, in response to Joseph I (Napoleon's brother), the contractual nature of royal authority and popular support. This also enabled the local elites to enjoy the authority they wanted. The societies over which they sought to preside were changing, demographically, economically, and politically, and this made the past a more problematic reference point.

Once returned to control in Spain, Ferdinand sought to reimpose it in Spanish America, sending an army in 1815, which, in effect, turned the revolution into a war by forcing forward counterrevolution. This attempt was initially successful. Royal authority was restored and the autonomy movements suppressed, although not that in the distant Plate (Plata) estuary. However, Ferdinand's cause faced many difficulties. The royalists in Latin America were badly divided, and their divisions interacted with contradictions within Spain's incoherent policies. Civil and military authorities clashed frequently, as did metropolitan and provincial administrations. Thus, in New Granada (now Colombia), the viceroy and the commander in chief were bitter rivals. Furthermore, financial shortages forced the royalist army to rely on the seizure of local supplies and on forced loans, which proved a heavy burden on the population and antagonized them from Spanish rule. The royalist forces sent from Spain were also hit hard by disease, especially yellow fever and dysentery, and were forced to recruit locally, leading to fresh political problems. New Granada had largely welcomed the royal army from Spain in 1815, but, by 1819, there was widespread support, instead, for an independent Colombia. Moreover, Spain did not possess any technological advantages akin to those enjoyed by the *conquistadores* in the early sixteenth century.

The situation was different in Mexico, where a revolutionary insurrection in 1810 had been hit hard in 1811–12 and 1815, while the insurrectionary guerrilla war had nearly ended by 1820. The royalist effort, however, was weakened by the liberal constitutional revolution in Spain in 1820, rather as the British effort in North America was undermined by the change in the government in 1782. Viewed as an unwelcome development by *criollo* (locally born) conservatives, and by those who wielded power in Mexico, this revolution led to a declaration of independence. In 1821, Augustín de Iturbide, the leading general, searching for a solution based on consensus, agreed with the rebels on a declaration of independence that proved widely acceptable. Under great pressure elsewhere, Spain accepted the situation that year.

Spain had proved unable to control the dynamic of events. Alongside persistent resistance within Spanish America, Spain repeatedly suffered from the willingness of Britain, the leading naval power, to provide assistance to the rebels, notably in the form of trade and recognition. By the end of 1825, Spanish control of the mainland was at an end. The empire was reduced to islands: the Canaries, Cuba, Puerto Rico, the Philippines, the Marianas, the Caroline islands, and the island-like enclaves of Spanish Morocco.

CIVIL WARS

In Spain itself, the overhang of the French Revolutionary and Napoleonic Wars was exacerbated by the political, economic, and fiscal crises linked to the loss of Spanish America. Ideological division was accompanied by conspiracies. Attempted coups by liberal military figures were defeated in Galicia in 1815 and Catalonia in 1817. Nevertheless, a liberal revolution in 1820, in which the royal palace was surrounded, obliged Ferdinand VII in 1821 to accept the 1812 constitution. The conservatives, however, rejected this and turned to violence, unsuccessfully mounting a coup in 1822, as well as encouraging a large-scale French military intervention in 1823 on behalf of Ferdinand.

In 1823, the opposing liberal Spanish army, short of supplies and unpaid, was affected by extensive desertion and retreated. Prefiguring the Spanish Civil War of 1936–39, the liberals were divided, while their anticlerical measures helped stir up popular antipathy, notably among the peasantry. Moreover, the liberal regime had become very unpopular in rural areas through its imposition of a cash economy, especially when it replaced tithes with cash payments that had to be calculated. Providing an opportunity for increasing financial demands on the peasantry, this measure contributed to the rapid collapse of the regime in 1823.

In the aftermath of Ferdinand's return to absolute power, liberals were executed, imprisoned, purged, or fled into exile. This, however, did not end the disorder. In 1827, the "Revolt of the Aggrieved" in the Catalan mountains was directed against those who were seen as Ferdinand's evil advisers. Support came from purged officers, but the uprising was swiftly suppressed. Yet again, Catalonia served as the basis for opposition to Castile.

Wars of revolution could play a formative role in Europe. In Italy, national unification in 1859–60, a step in which revolutionary activity in the south played a major role, was closely related to intervention by a leading Italian state—Piedmont—and by its non-Italian ally, France. At the same time, a nationalist revolution in Sicily and southern Italy proved crucial. Moreover, the entire conflict can be seen as a war that amounted to a revolution both within Italy and concerning its place in Europe.

Wars of revolution were civil wars, as sometimes, for those countries exposed to the pressure, were wars of imperialism. This was because the latter could entail significant division over whether to resist foreign conquest and rule. The wars of revolution that were civil wars were not invariably radical in direction. Instead, they could also be reactionary. In Spain, in the First Carlist War of 1833–40, dynasticism was a key context. Don Carlos, "Carlos V," resisted the bequest of the Spanish throne to his young niece, Isabel II, by her father and his brother, Ferdinand VII. Opposition to a female monarch was combined with hostility to the constitutional reform supported by Isabel's supporters and, more generally, to liberalism. Dynastism was impacted in specific social and regional circumstances. Carlism was a conservative movement that drew on peasant anger against liberal government, and thus reflected tensions that looked back not only to the Napoleonic period but also to opposition to the Enlightenment reforms of the late eighteenth century. Other Carlist wars followed in 1846–49 and 1872–76.

THE AMERICAN CIVIL WAR, 1861–65

The start of this war, the most traumatic in American history, is clear. On April 12, 1861, Confederate forces opened fire on Fort Sumter, the vulnerable federal position in Charleston harbor. The next day, the beleaguered fort surrendered after more than 3,000 shells and shot had been fired, setting fire to the wooden buildings in the fort. How the two sides got there and how, from that opening bombardment, they moved to war, however, is far more complex. Moreover, the issue of cause subsequently became involved in America's "history wars"; in 2000, when Representative Jesse Jackson Jr. and some other members of Congress claimed that at battle sites there was often "missing vital information about the role that the institution of slavery played in causing the American Civil War."[43]

Focusing first on 1861, Fort Sumter was not an accidental cause. President Abraham Lincoln had refused to yield to demands for its surrender. The fort represented the reality of national power in the face of a forge of Southern consciousness and separatism. The attack had been intended to intimidate him into yielding but, instead, led Lincoln to determine to act against what he termed "combinations" in the South. He went to war to maintain the Union, and not for the emancipation of the slaves. Lincoln's call for 75,000 volunteers, and his clear intention to resist secession with force by invading the lower South, played the major role in leading Arkansas, North Carolina, Tennessee, and Virginia to join the Confederacy; they did not intend to provide troops to put down what Lincoln termed an insurrection.

At the same time, it is necessary to row back to look at the causes of the war in a longer context. The tensions latent both in a federal system and in the diversity of the United States were the key points. This diversity was not only a matter of slave as opposed to free states, but that issue focused it. The admission of California as a free state in 1850 gave the free states a majority in the Senate, and the minority status of the South in the Union was a key feature of the sectional controversy of the 1850s, a feature that created problems for the South. A sense of being under challenge ensured that Southern secession was frequently threatened in the 1850s, before it formally triumphed in 1860–61. Tension in the South was raised greatly by John Brown's seizure of the federal arsenal at Harpers Ferry on the night of October 16, 1859. Intended to terrorize slaveholders and as the first stage of a war on slavery, to be achieved in large part by armed slaves, this rising by twenty-one men, however, was rapidly suppressed by Colonel Robert E. Lee and a force of Marines who stormed the building on the morning of October 18. Brown's small-scale uprising was scarcely a war, but it served, like earlier fighting in Kansas, to cross the boundary between contention and conflict. The stakes appeared raised as far as both the ends and the means of politics were concerned. Indeed, the significance of the uprising helps underline the difficulties, referred to in chapter 1, concerning how best to define war and, in particular, distinguish it from civil disturbances or political violence.

Many Southerners were convinced that the uprising revealed the true intentions of abolitionists, and this view proved a troubling background to the election campaign in 1860, as well as helping to accentuate the regional character of the contest. Indeed, the British envoy, Lord Lyons, reported, on December 12: "after making due allowance for the tendency to consider the 'present' crisis as always the most serious that has ever occurred, I am inclined to think that North and South have never been so near a breach."[44]

The alternative to secession was to seek to make the Union safe for the South and slavery, in part by reeducating Northerners about the constitution, or by acquiring more slave states, or by somehow addressing the vulnerabilities of the slave system in the South. The Southerners' failure to do so was compounded by the difficulties posed by the slaves' desire for freedom, although the latter did not have a scale nor disruptive consequences comparable to the situation in Brazil in the 1880s.

The 1860 election gave victory to Lincoln, who wished to prevent the extension of slavery into the federal Territories and who understood that this would threaten Southern interests and identity. The election of Lincoln, "a man almost unknown, a rough Western, of the lowest origin and little education,"[45] reflected the refashioning of politics by the slavery issue. First, there had been pressure on the Whigs, as with the rise of Libertyites and the Free Soil Party;[46] subsequently had come the disintegration of the Whigs in the aftermath of the Kansas-Nebraska Act of 1854, and the related rise of the Republicans as a Northern sectional party focused on the restriction of slavery. Lastly had come the division of the Democratic Party, "the only remaining bond between north and south,"[47] between Northern and Southern wings. In 1856, James Buchanan, the Democratic candidate, won the South but also Northern states, showing a national appeal. In contrast, Stephen Douglas, the Northern Democrat, competed with John Breckinridge, his Southern counterpart, in 1860. This competition allowed Lincoln, who carried none of the Southern states, to win on fewer than 40 percent of the votes cast.

National politics were no longer being contested by effective national parties and, partly as a result, American mass democracy could not generate a consensus. Compromise was certainly on offer, but it no longer seemed sufficiently acceptable to enough influential people in the North and South to gather impetus. Lincoln rejected a proposal by Senator John Crittenden of Kentucky that the 36°30′ line of the 1820 Missouri Compromise be run toward the Pacific, a line that would include the New Mexico Territory in the world of slavery. Slavery was a moral and cultural issue as well as an economic one, and that range helped both to cause the crisis and to lessen any chance of compromise.

Lincoln's election led to the secession of the South, beginning with South Carolina on December 20, 1860, and the formation of the Confederate States of America. That, however, did not equate with the slave states. Much of the Upper South had voted for John Bell of Tennessee, the candidate of the new Constitutional Union party pledged to back "the Union, the Constitution and the Laws," rather than for Breckinridge. Although Lincoln was willing to back a constitutional amendment prohibiting the federal government from interfering with slavery, nevertheless, secession was unacceptable to him and the Republicans. They argued both that the maintenance of the Union was essential to the purpose of America as well as to its strength and that it was necessary to understand that the superiority of the federal government over the states was critical to the idea of the American nation. Crucially, he turned to force in response to secession to limit its spread.[48] Choices were made against a background of panic. For example, fears of an abolitionist plot in Texas in late 1860, a major instance of the sequence of panics following Harpers Ferry, helped lead there to vigilante action and encouraged backing for secession.[49] Yet, the bombardment of Fort Sumter was crucial. This was action that could not be overlooked. It was an act of civil war by an official body in the shape of an army.

CONCLUSIONS

The variety of conflicts in the long nineteenth century is striking. So also was their ability to deliver verdicts. Given this, it was scarcely surprising that so many governments and movements chose to turn to war to further their interests. Moreover, fighting matched normative assumptions about the social and individual value of combat and military service. The values were sustained in cultures that were reverential of the past and referential to it. At the same time, the context was changing radically. The technological character of warfare was transformed more rapidly and comprehensively than before. Moreover, industrialization produced the resources for a total change in scale. In many countries, there were also new sociopolitical practices and structures that were linked to the new economics. These changed the parameters for governmental control, political consent, and, therefore, policy. The process and consequences varied by state, but it was significant not least in increasing the scale of government.

In general, it became less a case of one individual in control, and, instead, more one of rulership by a group. As a consequence, the group dynamics and psychological factors involved in war changed. This was not simply a new iteration of older patterns of Crown and aristocracy, although that was involved. In the new governmental systems, the military was a structured and coherent formal organization with a greater role in policy. The collective psychology of the elite in many, but not all, states was affected by this role, which often extended to the wider politics of the state. In this context, there was a tendency to envisage military outcomes to international alignments, let alone disputes, a tendency that drew on influential ideas of the inherent competitiveness of human societies and on survival and growth through strength and conflict. There was also a drive for role fulfillment in the shape of action, a drive that was to be crucial to the international crisis that led to the outbreak of World War I in 1914.

NOTES

1. Tim Blanning, *The Origins of the French Revolutionary Wars* (Harlow: Longman, 1986); Jeremy Black, *British Foreign Policy in an Age of Revolutions, 1783–1793* (Cambridge: Cambridge University Press, 1994), 335–545.

2. Michael Broers, *Napoleon: The Spirit of the Age, 1805–1810* (London: Faber and Faber, 2018); Adam Zamoyski, *Napoleon: The Man Behind the Myth* (London: William Collins, 2018).

3. John Tone, *The Fatal Knot: The Guerrilla War in Navarre and the Defeat of Napoleon* (Chapel Hill: University of North Carolina Press, 1994).

4. Lawrence Kaplan, "France and Madison's Decision for War, 1812," *Mississippi Valley Historical Review* 50 (1964): 652–71.

5. Threlkeld to John Fisher, March 1, 1817, DRO 1148M/19/9.

6. Richard Glover, "The French Fleet, 1807–1814: Britain's Problem; and Madison's Opportunity," *Journal of Modern History* 39 (1967): 249–51.

7. Melville to Popham, August 4, 1812, BL Loan 57/108. 114.

8. Paul Schroeder, *The Transformation of European Politics, 1763–1848* (Oxford: Oxford University Press, 1994), 435–40, quote: 440.

9. Boyd Hilton, *A Mad, Bad, and Dangerous People? England 1783–1846* (Oxford: Oxford University Press, 2006), 220.

10. Smith to William Pinkney, Minister in London, May 22, 1810, *Documents Accompanying the Message of the President* (Washington, 1810), copy in BL Add. 49178 fol. 3.

11. Bradford Perkins, *Prologue to War: England and the United States, 1805–1812* (Berkeley: University of California Press, 1961), 55.

12. *Annual Register* 54 (London, 1812): 420–23.

13. Lance Davis and Stanley Engerman, *Naval Blockades in Peace and War: An Economic History since 1750* (New York: Cambridge University Press, 2006).

14. BL Add. 49990 fol. 25.

15. BL Add. 49990 fols. 17–20.

16. H. W. Fritz, "The War Hawks of 1812," *Capitol Studies* 5 (Spring 1977): 25–42.

17. A. DeConde, *The Quasi-War: The Politics and Diplomacy of the Undeclared War with France, 1797–1801* (New York: Scribner's, 1966).

18. B. H. Tolley, "The Liverpool Campaign against the Order in Council and the War of 1812," in *Liverpool and Merseyside: Essays in the Economic and Social History of the Port and its Hinterland*, ed. J. R. Harris (London: Frank Cass, 1969); D. Moss, "Birmingham and the Campaigns against the Orders-in-Council and East India Company Charter, 1812–13," *Canadian Journal of History* 11 (1976): 173–88.

19. Castlereagh to Lords of the Admiralty, August 6, 1812; John Croker, Secretary to the Admiralty, to Warren, August 7, 1812, NA ADM 1/4222, 2/1735.

20. Isabel Kelsay, *Joseph Brant, 1743–1807: Man of Two Worlds* (Syracuse, NY: Syracuse University Press, 1984); Wiley Sword, *President Washington's Indian War: The Struggle for the Old Northwest, 1790–1795* (Norman: University of Oklahoma Press, 1985).

21. Peter Marshall, "Britain Without America—A Second Empire?" in *The Oxford History of the British Empire. II. The Eighteenth Century*, ed. P. J. Marshall (Oxford: Oxford University Press, 1998), 579.

22. Richard White, *The Middle Ground: Indians, Empires, and Republics in the Great Lakes Region, 1650–1815* (Cambridge: Cambridge University Press, 1991); G. E. Dowd, *A Spirited Resistance: The North American Indian Struggle for Unity, 1745–1815* (Baltimore, MD: Johns Hopkins University Press, 1992); K. DuVal, *The Native Ground: Indians and Colonists in the Heart of the Continent* (Philadelphia: University of Pennsylvania Press, 2006).

23. BL Add. 49990 fols. 59–60.

24. R. M. Owens, "Jeffersonian Benevolence on the Ground: The Indian Land Cession Treaties of William Henry Harrison," *Journal of the Early Republic* 22 (2002): 434.

25. G. E. Dowd, "Thinking and Believing: Nativism and Unity in the Ages of Pontiac and Tecumseh," *American Indian Quarterly* 16 (1992): 324–26.

26. J. F. Hopkins, ed., *The Papers of Henry Clay* (Lexington: University of Kentucky Press, 1959), 1:642.

27. Jefferson to Adams, June 11, 1812, *The Adams-Jefferson Letters*, ed. L. J. Cappon (Chapel Hill: University of North Carolina Press, 1959), 308.

28. R. Kagan, *Dangerous Nation: America and the World, 1600–1898* (London: Atlantic, 2006), 145.

29. For a report of September 1814 to the British government on American views on these lines, BL Add. 38259 fols. 93–94.

30. *Clay*, 1:842.

31. R. Horsman, "Western War Aims, 1811–1812," *Indiana Magazine of History* 53 (1957): 1–18; W. R. Barlow, "The Coming of the War of 1812 in Michigan Territory," *Michigan Territory* 53 (1969): 91–107.

32. Winfried Baumgart, *The Crimean War, 1853–1856* (London: Arnold, 1999).

33. Joseph Smith, *The Spanish-American War: Conflict in the Caribbean and the Pacific, 1895–1902* (Harlow: Longman, 1994).

34. I. R. Smith, *The Origins of the South African War, 1899–1902* (Harlow: Longman, 1996).

35. Ian Nish, *The Origins of the Russo-Japanese War* (Harlow: Longman, 1996).

36. B. Williams, "Approach to the Second Afghan War: Central Asia during the Great Eastern Crisis, 1875–1878," *International History Review* 2 (1980): 216–17.

37. J. Christopher Herold, *Bonaparte in Egypt* (London: Hamish Hamilton, 1962), 3–4.

38. George, 4th Earl of Aberdeen, Foreign Secretary, to Richard Pakenham, envoy in Mexico, 18 May 1844, NA FO 5/403 fol. 53.

39. Waddy Thompson, *Recollections of Mexico* (New York, 1847; reprint, Creative Media Partners, 2015), 170.

40. Pakenham to Aberdeen, May 28, 1846, BL Add. 43123 fol. 295.

41. Pakenham to Aberdeen, June 23, 1846, Southampton, University Library, Palmerston papers, BD/US/51.

42. BL Add. 49968 fol. 8.

43. Dwight Pitcaithley, "'A Cosmic Threat': The National Park Service Addresses the Causes of the American Civil War," in *Slavery and Public History: The Tough Stuff of American Memory*, ed. James Horton and Lois Horton (New York: New Press, 2006), 169–86; Robert Cook, *Troubled Commemoration: The American Civil War Centennial, 1961–1965* (Baton Rouge: Louisiana State University Press, 2007).

44. Lyons to Earl Russell, Foreign Secretary, December 12, 1859, NA PRO 30/22/34 fol. 69. See, more generally, Jason Phillips, *Looming Civil War: How Nineteenth-Century Americans Imagined the Future* (Oxford: Oxford University Press, 2018).

45. Lyons to Russell, May 22, 1860, NA PRO 30/22/34 fol. 150.

46. Bruce Laurie, *Beyond Garrison: Antislavery and Social Reform* (Cambridge: Cambridge University Press, 2005).

47. Lyons to Russell, April 30, 1860, NA PRO 30/22/34 fol. 136.

48. Russell McClintoch, *Lincoln and the Decision for War: The Northern Response to Secession* (Chapel Hill: University of North Carolina Press, 2008).

49. Donald Reynolds, *Texas Terror: The Slave Insurrection Panic of 1860 and the Secession of the Lower South* (Baton Rouge: Louisiana State University Press, 2007).

Chapter Six

World War I and Its Sequel, 1914–30

PRESSURES FOR WAR

The causes of World War I recur in all major academic and popular studies of the causes of war. They are doubly significant as a topic in that this is the war that also features in most public discussion of the causes of war, a result encouraged by extensive consideration at the time of the centenary in 2014. That situation, however, does not necessarily help understanding.

At the outset, it is helpful to focus on the attitudes of 1914. The prospect of major war was discussed and planned for by many in the decades prior to the outbreak of World War I in August 1914, but the probable nature and, still more, consequences of the resulting conflict were understood by few. War seemed likely because both experience and assumptions led in that direction. In the previous century, key issues were settled by conflict, whether the two overthrows of Napoleon I of France by a European alliance (1814 and 1815), the Unification of Italy (1860), the eventual maintenance of the American union in the face of the Civil War (1861–65), the transformation of Prussia into the German empire thanks to repeated triumphs in the Wars of German Unification (1864, 1866, 1870–71), or the rise of Japan, with victories over China (1894–95) and Russia (1904–5). This process was also true for other, lesser, states such as Italy, Serbia, Greece, Romania, Bulgaria, and Ethiopia, all of which traced their independence and expansion to success in recent warfare.

Conversely, states that had failed in such warfare, for example the Ottoman Empire (Turkey for short) or Bulgaria, both of which had been heavily defeated in the conflicts of 1911–13, the Balkan Wars, saw such defeats as an encouragement to reverse failure through subsequent struggles. Movements that lacked statehood, such as Irish and Polish nationalism, also looked to past defeats, notably unsuccessful rebellions against British and Russian rule in 1798 and 1863 respectively, as a call for fresh valor. These movements (correctly) saw war between the major states as an opportunity to press their claims. So also did second-rank states.

War and victories as a measure of national success constituted a key ideological and practical predisposition to struggle. This predisposition had a variety of bases, including the intellectual conviction that such struggle was a central and inevitable feature of natural and human existence and development, as well as a cultural belief that struggle expressed and secured masculinity, and thus kept both society and civilization vital. This was a view, for example, that both the nationalists and the Futurists, a cultural movement, could share in Italy. Doubt was presented as female and associated with weak emotions and feelings. Belief in war, as an expression of a martial spirit and an ideology of masculinity, was greatly sustained by the popular literature in Europe, the United States, and Japan. In Britain, those who volunteered to fight in 1914–16 were the generation who would have grown up in the 1890s and 1900s reading storybooks like *Union Jack*, *Captain*, and *Chums* and also the novels of G. A. Henty. Although Britain was not bellicose as far as other European powers were concerned, these and other publications had promoted popular militarism, especially in an imperial context. [1]

Linked, for some, to these views was a sense of anxiety based on belief that the present situation was necessarily unstable and also prone to decline, decay, and degeneration. Such a fate apparently could only be avoided by vigilance, effort, and sacrifice. This cultural anxiety was accentuated by concerns about the alleged consequences of industrial society, urban living, and democratic populism. These concerns were focused in some cases, notably Germany, by an opposition to the left-wing politics believed to flow from these developments. Socialism was seen on the Right as a threat to Germany's ability and willingness to fight, and a similar view was held elsewhere.

There were also doubts about the strength of masculinity in the face of cultural and social changes. These doubts were related to worries about national degeneration in a context of a belief in a Darwinian competition between nations and races, a competition that was seen as inherently violent. [2] This approach in fact rested on a false understanding and a corruption of Darwinian theory from "Survival of the Fittest," which, as originally conceived, did not apply to states or peoples or groups within society, but to species and adaptations within species, for which the dynamics were different. Thus, Darwin's arguments were misused to justify aggression and domination, and to offer a misplaced clarity, one in which entitlement was presented as need.

These factors were accentuated by the apparent exigencies of an international system in which the only choice seemed to be between growth and decay, empire and impotence, standing by allies or showing weakness. [3] To fail to act was, allegedly, to be doomed to failure. This is an assumption that has frequently proved effective in gaining support for war. Furthermore, the territorial expansionism of the imperialism of the period gave a tone of greater competition to international relations, with alliances of mutual restraint being replaced by alliances focused on securing additional power.[4] Expansionism also rose from a belief that there was a simple choice of growth or decline. This belief encouraged a concern with relative position. Repeatedly, a perception of the reality, or potential, of relative loss was crucially important in helping center anxieties, and a need for action, in particular conjunctures. Existing pacts, such as the Triple Alliance of Austria, Germany, and Italy in 1882, pacts that had often restrained by joining powers with different interests, notably Austria and Italy, appeared inadequate in the 1900s and 1910s as anxieties grew about shifts in international geopolitics and national politics. This point serves as a reminder of the difficulties of operating deterrence and also of the extent to which the international architecture can have different, indeed very different, meanings and values depending in part on the ideological context of the moment.

The imperial hopes, dreams, anticipations, expectations, anxieties, and nightmares of the great European powers affected, and were affected by, the fate both of the non-European world and that of southeastern Europe, and to a degree not seen before. In particular, anxieties in, and about, the Balkans were to press directly and strongly on competing European alliance systems, ensuring that the limited wars and compromises at the expense of others that were seen with large-scale European expansionism outside the continent could not eventually be sustained as a system within Europe itself.

These factors encouraged bellicosity, or at least an acceptance that war might be noble and strengthening as well as necessary; but, on a continuing pattern, they did not explain why large-scale conflict seemed more of a risk in the 1900s and early 1910s than earlier, or why it broke out in 1914, and not earlier. Such explanation in part rests on contrasting assumptions. In particular, there is a tension between a "systemic" account of the outbreak of war, which would trace it to the nature of the competitive international system, or, alternatively, one that places greater weight on the agency (actions) of particular powers. The latter is a more convincing account; it makes greater allowance for the extent to which individual leaders and specific policy-making groups took the key decisions, including the crucial decisions of how, and when, to act. Insofar as a systems approach is adopted, this account is one that offers more for the role of strategic cultures and for the mismatch between them and the disputes and uncertainties that might arise as a result.

In particular, it is apparent that the leading factor was the encouragement provided to Austria (short for Austria-Hungary, the empire ruled by the Habsburgs) by German policy makers, notably the Kaiser [Emperor], Wilhelm II (r. 1888–1918), and his military advisers. The nervy Kaiser was volatile, weak, and highly competitive, not least toward Britain: he was hostile to his British mother.[5] The conservative German elite was worried about domestic changes, including left-wing activism, as well as by international challenges. Reflecting the atavistic roots of much militarism and imperialism, and the prejudices of a traditional elite who felt threatened by change and modernization, and who used militarism to entrench, as well as reflect, their privileges, the German regime, like others, was operating in an increasingly volatile situation, in which urbanization, mass literacy, industrialization, secularization, and nationalism were creating an uncertain and unfamiliar world. There was a particular spatial dynamic in Germany in that the wealth and political activism of industrial advance was concentrated in the west, notably the heavy industry of the Ruhr valley and the oceanic trading center of Hamburg, but much of the military elite based their position on landholdings in the agrarian east. Their relative value had diminished both within Germany and, as a result of increased transoceanic trade, globally. Moreover, those who dominated the agrarian east felt threatened by Russian strength in neighboring Poland.

Faced by international and domestic challenges, the temptation, both in Germany and Austria, was to respond with force, to impose order on the flux, or to gain order through coercion. Germany was affected domestically by pressures for welfare and education reform, and by religious divisions. Militarism, in contrast, appeared to offer a source of unity and agreement. A growing sense of instability both encouraged the use of might to resist or channel it and provided opportunities to do so. In large part, militarism fed itself in Germany, particularly after the failure of the liberal revolutions there in 1848, for, instead, German nationalism was focused on the more authoritarian political culture of Prussia. By means of defeating Austria in 1866 and France in 1870–71, Prussia, which had earlier been one of the weaker and, due to its central position, more vulnerable of the major powers, created the German empire, and the Hohenzollern kings became emperors.

German strength in a central position in Europe thereby created a new set of geopolitical tensions—one that was unprecedented since the failure of the Habsburgs to dominate Germany, a failure that became clearly apparent in the mid-eighteenth century. In part, tensions arose because the annexation of most of Alsace and part of coal- and iron-rich Lorraine from France as part of the peace settlement in 1871 left—as Otto von Bismarck, the German Chancellor, had been warned by the French emissary—a long-term sense of French grievance that posed a lasting security challenge to Germany on its western border.[6] This challenge was greatly mitigated, however, by Germany's far larger population and, therefore, army, and her stronger economy. France, moreover, was weakened by its competition with Britain in imperial expansion, especially, but not only, in sub-Saharan Africa. The two powers nearly came to war over control of Sudan in the Fashoda Crisis of 1898.

Bellicosity in that case, however, was lanced by the possibilities of compromise at the expense of the vastness of Africa. In return for France's acceptance of Britain's position in Egypt and Sudan, Britain was to accept French expansion in Morocco.

Instead of fearing France alone, German policy makers saw danger in the possibility of French cooperation with other powers overshadowed by Germany's rise to power. This was cooperation with Austria, defeated in 1866, and Russia, which, without any conflict, had lost relative strength with this rise. Bismarck, one of the key figures in German unification, secured his achievement by means from 1871 of alliance with Russia and Austria, notably the *Dreikaiserbund* (League of Three Emperors) of 1881, which was a reincarnation of Tsar Alexander's Holy Alliance of 1819, one that had a political and ideological weight alongside its international significance.

However, this achievement was neglected by his successors, notably because the arrogant Wilhelm II, who parted company with Bismarck in 1890, was scarcely risk-averse and, instead, was committed to expansion, not stability. Wilhelm also viewed Russia with suspicion and regarded it as a racial threat. The non-renewal of the German-Russian Reassurance Treaty represented a clear strategic deviation from Bismarck's foreign policy and his overriding strategy of keeping France isolated, and thus deterring her from seeking to overturn the system.

Anxiety was a key element throughout, but the process of anxiety-driven aggression was particularly apparent with both Austria and Germany. The Austrian elite worried that the breakdown of Turkish power, notably as a result of Turkey's heavy defeat by the Balkan powers in the First Balkan War of 1912–13, was leading to a degree of nationalist assertiveness in the Balkans, especially by Serbia, which made major territorial gains as a result of the war. This assertiveness threatened the cohesion of the Austrian empire, not least because Serbia encouraged opposition to Austrian rule in neighboring Bosnia, which Austria had occupied in 1878 and annexed in 1908. Whether Austria was itself overstretched in the face of rising nationalism within the empire, which included the modern states of Croatia, Slovenia, Bosnia, Slovakia and the Czech Republic, and parts of Poland, Romania, and Italy, is a matter for controversy. However, political disputes related to this nationalism, some of which was separatist in character, affected the ability to pursue policy initiatives. Moreover, these disputes helped create a destabilizing sense of the enemy within, with these enemies having friends outside the empire.

In Austrian eyes, this enemy would be neutered by reordering the Balkans in a way that served Austria's interests and demonstrated its superiority, and specifically by weakening Serbia. Ironically, an equal, or more serious, danger to the Austrian empire came not from Slav nationalism, but from a bellicose nationalism on the part of much of the ruling Austrian elite. There was both a misleading belief in military solutions to internal and external problems and a serious overestimation of their military potential. This bellicosity accentuated Austrian domestic and international weaknesses. It was not the threat but the response that was the key element, and this response was a matter of both content and tone. Again, this is a frequent issue in the causes of war.

In Germany, there was concern about the problems of its Austrian ally and, as a result, of the problems of having Austria as an ally, although not to a sufficient degree. In the event, neither Germany nor Austria was to restrain the other sufficiently in 1914. This proved the key instance of the way that, with this war, the geopolitical logic of alliances drew powers into actions that were highly damaging and, in the end, destroyed the logic of the alliances. Austria believed in its own might and capability, and in its German allies in case anything went wrong.

Yet, although Austria helped cause the war, Austria was not the crucial element for Germany, for there was fear of Russians, not Serbs, in Berlin, and not of Russians and Serbs as there was in Vienna. The negotiation of a Franco-Russian military agreement in 1894 had led Germany, fearing, indeed predicting, war on two fronts at once, to plan to achieve by speed the sequential war-making (fighting enemies one after another) that had brought victory to Prussia in 1864–71 over Denmark, Austria, and France, and victory to Napoleon I of France in 1805–7 over Austria, Prussia, and Russia.

However, it appeared difficult for Germany to achieve a knockout blow against Russia, the frontier of which had moved west to include much of Poland as a result of the major role of the Russians in victory over Napoleon I in 1812–14. This westward movement, a key geopolitical factor, provided the defense of depth which indeed was to enable Russia to survive considerable territorial losses to Germany in the first three years of the Great War and again in 1941. This defense of depth led German planners to focus prewar, first, on defeating France before turning against Russia. Unlike Moscow, Paris appeared to be within reach of German forces, as had been shown unsuccessfully in 1792, and successfully in 1814 and 1870, and was to be nearly shown again in 1914, 1918 and, finally, successfully, 1940.

This strategy depended, in 1914, on a key capability gap that had uncertain operational results and strategic consequences: the contrast between rapid German mobilization and its slower Russian counterpart. As a result of the significance of this gap, there was German concern as Russian offensive capability improved. This was a capability financed by France, not least in the shape of railway construction in Russian Poland designed to speed the movement of Russian troops westward toward the Austrian and German borders.[7] Each power had benefited from the three partitions of Poland in 1772–95 in order to seize all of the country between them. The Russian military program announced in 1913 caused particular anxiety in Berlin, which gave Count Helmuth von Moltke (Moltke the Younger, nephew of the Moltke of the Wars of Unification), chief of the German General Staff from 1906, further motivation in 1913–14 to press for war. He feared that Germany would increasingly not be able to win later (a verdict endorsed in very different circumstances by World War II) and, by 1916, he thought, Russia, with its larger population, would be in a position to start attacking effectively before the Germans had had an opportunity to defeat France. In Moltke's view, Germany's existing plans could not operate after 1916.

Had Moltke been open to other ways of thinking, matters might have been different. Ironically, although the Russians had built up an army superior to that of Austria, against whom they repeatedly did well in 1914–16, and had quickly recovered from defeat by Japan in 1904–5, the Russian attack on Germany was to be defeated easily and rapidly, and by only a small section of the German army, in 1914. Moreover, much of Russian Poland was conquered by the Germans in 1915 even though Germany was then fighting a two-front war and under attack on the Western Front.

The Germans, both military and politicians, had consistently overestimated Russia's military potential, in part because they exaggerated the quantitative indices of army strength at the expense of qualitative criteria. This factor indicates the more general problems of assessing strength, which is a major problem with the theory and practice of deterrence. These problems with assessment are more generally true of planning. Numbers are fixed and easily calculated unlike motivation and other non-metric factors that in practice play a key role.

There is nothing to suggest that the German overestimation of Russian strength was deliberate, but the misperceptions proved very powerful. They owed much to more widespread irrational fears that overshadowed practicalities and that also satisfied aspirations and concerns focused on promoting war. Again, these are more generally relevant points.

The defeat of France, in contrast, was regarded as probable because the Germans, encouraged by the example of victory over France in 1870–71, assumed that their better-prepared forces would win regardless of French actions. Although Moltke knew that the German army was not inherently superior to that of France, German commanders were very much in the shadow of expectations created by the repeated successes of 1864–71, just as British naval commanders were greatly affected by the Nelsonian legacy in their quest for a supposedly decisive, and certainly glorious, victory. Affected by advances in technology and planning methods, German commanders felt that their plan would work even better than those of 1864–71. The majority and, after 1906, all the British service attachés in Berlin reported that the German armed forces were preparing for attack and 1913–1915 was seen as the likeliest period for this aggression.[8]

Moltke's predecessor, Count Alfred von Schlieffen, had changed the General Staff when he was its head from 1891 to 1906, allowing its members to become military specialists at the expense of more general, nonmilitary, knowledge. Nonmilitary problems were consciously excluded from General Staff thinking. Military decision makers therefore were allowed to conduct their planning in a vacuum, with scant regard for the political situation around them. In 1914, by stressing future threats and affirming that Germany was still able to defeat likely opponents, the General Staff helped to push civilian policy makers toward war—although, in practice, civilians were not greatly consulted by the military, which had independent access to Wilhelm II. Moltke hoped that the resulting conflict would be the short and manageable war for which the Germans had been planning, but he feared that it could well be a long war, a struggle indeed for which Germany was unprepared. Because Moltke did not develop an alternative plan (instead deciding to scrap one in 1913), the option of deploying troops only in the East against Russia no longer existed in 1914.[9] Planning focused on France, and neither Germany nor Austria prepared adequately for war with Russia, which was an aspect both of how poorly they worked together and of the short-term nature of most military planning.

German Intelligence capacity proved erratic, as indeed it also did in World War II. In 1941, there was to be inadequate planning by Germany for war with the Soviet Union, and in 1942 for the renewed attack. Inadequate planning was matched by poor preparation. In both world wars, the emphasis on "rational" planning encouraged the setting of goals that appeared viable but were over-ambitious.

German policy makers believed that they could control the situation created by their Austrian ally, that if war came they had the necessary plan, and that their army was the best and capable of executing the plan. Alongside anxiety about Russia, there was an ambition on the part of German leaders that was more clearly a case of wishful thinking and strategic overreach, a process also seen in the following world war. This ambition in 1914 was for becoming not simply a great power but a world power, able to match Britain in this, and thus to overthrow and replace her imperial position. To do so, Germany sought a navy able to contest that of Britain. This drive, in practice, was unnecessary to Germany's goals within Europe. Moreover, naval ambition was likely to alienate Britain seriously, both the government and the public, and, therefore, to ensure that these goals became unattainable, as indeed was to be the case. The assumptions that Britain's differences with France and Russia would remain inseparable, and thus would lessen Britain's concern about Germany and make alliance with these powers unlikely, proved greatly mistaken. Moreover, the Japanese elimination of Russia as a naval power in 1905, notably at the battle of Tsushima, undermined Germany's calculations as it helped the British to focus on the German challenge rather than, as in the 1890s, those of France and Russia.

Fear of German intentions, and particularly of her naval ambitions, encouraged closer British relations with France from 1904. The Anglo-French entente of that year led to military talks between Britain and France in part because defeat in the Russo-Japanese war of 1904–5 weakened Russia (France's ally) as a balancing element within Europe, thereby exposing France to German diplomatic pressure, and also creating British alarm about German intentions, as in the First Moroccan Crisis of 1905–6. This crisis, provoked by Germany and an instance of how a struggle for primacy in a peripheral area could lead to warlike moves—a lesson for today—was followed by Anglo-French staff talks aimed at dealing with a German threat. In 1907, British military maneuvers were conducted on the basis that Germany, not France, was the enemy. Moreover, also that year, fears of Germany contributed to an Anglo-Russian entente that eased tensions between the two powers, notably competing ambitions and contrasting anxieties in South Asia. Germany, with its great economic strength, its naval ambitions, and its search for a "Place in the Sun," was increasingly seen in Britain as the principal threat. This was not an unreasonable view, as such a drive would certainly have followed German mastery of the Continent, again a situation that anticipated that in the following world war.

The economic statistics were all too present to British commentators, not least because they enabled Germany to pursue its naval race for battleship strength with Britain from 1906. The annual average output of coal and lignite in million metric tons in 1870–74 was 123 for Britain and 41 for Germany, but, by 1910–14, the figures were 274 to 247. For pig iron, the annual figures changed from 7.9 and 2.7 in 1880 to 10.2 and 14.8 in 1910; for steel from 3.6 and 2.2 in 1890 to 6.5 and 13.7 in 1910. In 1900, the German population was 56.4 million, but that of Britain excluding Ireland only 37 million and including her still only 41.5 million.

In December 1899, the rising journalist J. L. Garvin, whose son was later to be killed in the war, decided that Germany and not, as he had previously thought, France and Russia, was the greatest threat to Britain. Rejecting the view of Joseph Chamberlain, Secretary of State for the Colonies, that Britain and Germany were natural allies, their peoples of a similar racial "character," Garvin saw "the Anglo-Saxons," in other words the British, as the obstacle to Germany's naval and commercial policy. Imaginative literature reflected, and contributed to, the sense of crisis. A projected German invasion of Britain was central to *The Riddle of the Sands* (1903), a popular adventure novel by Erskine Childers that was first planned in 1897, when indeed the Germans discussed such a project.

Yet, political opinion was divided. There were influential British politicians who sought to maintain good relations with Germany. Moreover, the ententes with France and Russia were not alliances, and Britain failed to make her position clear, thus encouraging Germany to hope that Britain would not act in the event of war, which was also Hitler's mistaken belief when he invaded Poland in 1939. In 1914, as a separate issue, the British certainly failed to make effective use of their fleet as a deterrent, restraining Germany from hostile acts. However, such a means and outcome were scarcely possible, in large part because of the German conviction that quick victory was feasible thanks to rapid success on land whatever threat the British fleet posed in the long term. This conviction proved totally misplaced.

In the Franco-Prussian War of 1870–71, Germany had won despite greater French naval strength. In any future struggle, if Britain remained neutral, Germany, whose navy was now bigger than that of France, could hope to trade both with her and with the United States, thus deriving an economic benefit that would make it easier to pursue her goals within Europe. Instead, the competitive naval race with Britain from 1906 was based largely on a feeling of inferiority toward her, in part arising from the serious psychological issues of the maladjusted Kaiser, whose mother was the daughter of Queen Victoria (r. 1837–1901), and who was somewhat obsessed by naval matters.

Nevertheless, despite this competition, there was a lack of effective German planning for a naval war with Britain, and the priority placed on the German army, notably with the Army Bill of 1913, was such that by 1914 naval tensions between the two states had lessened. The concern of Theobald von Bethmann-Hollweg, who became chancellor in 1909, about naval costs was matched by the army's emphasis on the needs of a two-front war on land. The commitments in the Army Bill of 1913 represented, in effect, a unilateral German declaration of naval arms limitation, albeit at a high level of competition and annual completion of warships. The Germans were also deterred by the pace of British naval shipbuilding, which, in direct response to Germany, increased greatly from 1908. [10]

Naval interests also led in a different strategic direction to those of the army, which reflected and sustained the seriously dysfunctional character of German interservice war preparations, one also to be seen with Japan in 1941. The German army and the navy had two completely different approaches. The army wanted to invade Belgium and leave Denmark alone, the policy followed in 1914, whereas the navy wanted precisely the opposite in order to leave Belgium and the Netherlands, especially the major port of Rotterdam, as a windpipe for German imports in anticipation of the British blockade of Germany, while strengthening the German position in the North and Baltic seas by conquering Denmark. There was no effort, on the part of Wilhelm II or anyone else, at the slightest semblance of joint planning, which would have been seen as an acknowledgment of army weakness. Indeed, at the beginning of the war, long-range economic planning was completely ignored in Germany, where it was treated as heresy against the confident belief in a short war. The passive nature of a blockade was not welcome.

European tensions rose in the years after 1910, with the respective alliances increasingly concerned about the real actions and supposed intentions of their rivals. For example, the visits of President Raymond Poincaré of France to Russia in 1912 and 1914 seemed to underline the apparent danger to Germany posed by their alliance, and therefore the need for protective action. The French increase in army spending in 1912, a response to the German move of the previous year, in turn led to a German rise in expenditure on the army. These increases and other moves encouraged a sense of instability and foreboding and helped drive forward an arms race on land. In turn, this greater military capability, by lessening earlier weaknesses, made armed diplomacy more plausible while, at the same time, increasing a sense of vulnerability to the armed diplomacy of others. As deterrence appeared weaker, so it seemed necessary to identify and grasp windows of opportunity for action, which, in turn, greatly weakened deterrence.

The Germans obtained further evidence of closer links between Britain, France, and Russia in May 1914, with espionage information on Anglo-Russian talks for a naval agreement. The Germans increasingly felt encircled and threatened, but, as a reminder that there were also failings in analysis elsewhere, British strategy was affected by an inability to grasp the consequences of closer diplomatic relations with France and Russia. In the sense that they, together, instead of serving as a deterrent, could make war with Germany more likely, this outcome was only appreciated to a degree, but it was certainly not sought.

Russia attempted to recover from its defeat by Japan by building up its military anew with resources that came from economic growth and French loans, notably that of 1905, which gave Russia access to French capital markets. The percentage of total Russian expenditure on the military rose from 23.2 in 1907 to 28.2 in 1913, although the majority of Russian officers remained unable to put men and equipment to good use and, indeed, the Russo-Japanese War led to a conservative reaction in Russian military circles against attempts to reform operational practice.

From 1911, when the prime minister, Peter [Pyotr] Stolypin, was assassinated, Russia was under a more interventionist and aggressive government, less willing to subordinate geopolitical goals to domestic issues. Russia became Serbia's protector, although they were not formally allies; the relationship was based upon Russia's implicit support for the Bulgarian and Serbian alliance of March 1912 and her explicit agreement to settle any ensuing disagreement over the disposition of the region of Macedonia when it was conquered from Turkey. There was a religious dimension; Russia, Bulgaria, and Serbia were all Orthodox states opposed alike to Islam and Catholicism, the latter the religion of the Austrian and Hungarian elites. In the autumn of 1912, international tension over Serbian policy led Austria and Russia to deploy troops in mutually threatening positions, but these forces withdrew in the spring of 1913.

The 1914 crisis, however, had a very different outcome. Visiting Sarajevo, the capital of the province of Bosnia, Archduke Franz Ferdinand, the nephew and heir to the very elderly Emperor Franz Joseph of Austria, and his wife Sophie, were assassinated on June 28, 1914, by Gavrilo Princip, a Bosnian Serb, leaving black-and-white photographs of violence in hot streets. The terrorist group was under the control of the Black Hand, a secret Serbian nationalist organization pledged to the overthrow of Austrian control in South Slav territories, notably Bosnia. "Apis," Colonel Dragutin Dimitrijević, the head of Serbian military intelligence, was a crucial figure, able to ignore his government's efforts to contain the activities of the Black Hand, which he had founded in 1911. Apis sought both to create a Greater Serbia and to overthrow the Serbian prime minister, Nikola Pašić. Indeed, Apis and two allies were to be executed in 1917 on charges of conspiring against the then Serbian government-in-exile.

When the news of the killings in Sarajevo reached Vienna, there was shock and the customary response, not least by a militaristic state, to an unexpected and dramatic event, a sense that a display of action and power was required. This sense interacted with an already powerful view that war with Serbia was necessary, and this new situation apparently provided the excuse to take care of Serbia, a policy already discussed in 1912 and 1913. Believing that German backing would deter Russia, an inaccurate assessment, the Austrians sought not agreement with the Serbs, who, in fact, were willing to make important concessions and prepared to accept binding arbitration on the points to which they objected, but a limited war with Serbia.

This policy was pushed by a key advocate of the value of force and the necessity of war, Franz von Hötzendorf (referred to as Conrad), the self-absorbed and seriously flawed head of the Austrian General Staff. In addition, aside from the conversion of Leopold Berchtold, the Foreign Minister, to a military solution to the challenge apparently posed by Serbia to Austrian rule, the aristocratic culture of the Austrian diplomatic corps did not favor compromise. Instead, the South Slavs were generally viewed with contempt, and there was a strong cultural preference supporting the alliance with Germany. At the same time, military decision makers, not diplomats, played the vital role in pushing for conflict with Serbia.[11] Alongside a number of other Austrian policy makers, Conrad believed that war was the best way to stabilize the Habsburg monarchy in the face of serious nationalist challenges from both within, notably Czechs and Poles, and without, especially Italy and Serbia. The killings at Sarajevo thus provided him with the opportunity to carry out his belief in preventive war.

German support to Austria meant crucial encouragement. This support reflected the belief in Berlin that a forceful response was necessary, appropriate, and likely to profit Austrian and German interests at a time when an opportunity for success existed. The Serb response of July 25 to a deliberately unacceptable Austrian ultimatum of July 23 was deemed inadequate and, on July 28, without pursuing the option of further negotiations, the Austrians declared war. The Russians, meanwhile, responded to the ultimatum to Serbia by beginning military preparations on the 26th. They were confident of French support and believed it necessary to act formally to protect Serbia.

On July 30, Russia declared general mobilization, a step that the Germans had already been preparing to take. Erich von Falkenhayn, the war minister, wanted to mobilize the previous day. As a result of Russia acting first, the Germans were able to present their step, in a misleading fashion, as defensive. Doing so helped lessen potential domestic opposition, notably in the *Reichstag*, to the voting of necessary credits for war. German mobilization was, in part, a question of timing in the knowledge that Russia was about to mobilize.

In line with its planning, but also with a paranoia about encirclement, the German military was convinced that it must win the race to mobilize effectively and to use the resulting strength, and notably before the Russians could act against both Germany and its ally Austria. This concern encouraged the Germans, if they could not use the crisis to divide France and Russia, to attack both. In part, they were led by their strategic concepts and operational concerns, notably how best to ensure victory in any war that broke out, but, throughout, German leaders opportunistically sought to use the Balkan crisis to change the balance of power in their favor. They were willing to risk a war because no other crisis was as likely or planned to produce a constellation of circumstances guaranteeing them the commitment of their main ally, Austria, and the support of the German public.

Thus, it is too much to say, as has been argued for Germany, that the war plans of 1914, with their dynamic interaction of mobilization and deployment, made "war by timetable" (a reference to the railway timetables that guided and registered the pace of mobilization) difficult to stop once a crisis occurred. Instead, such an argument both exaggerates the role of one particular factor and underplays the extent to which rulers, generals, and politicians were not trapped by circumstances. In fact, as was understood, their own roles, preferences, and choices were important. An underplaying of the importance of choice reflects an anachronistic, later, sense that no one could have chosen to begin the war; but, in fact, in 1914, decision makers believed that war was necessary and could lead to a quick victory. Such an underplaying also permits the convenient blaming of "the system," as opposed to individuals who were foolish, overconfident, or weak.

The role of choice is illustrated by the extent to which an awareness of likely risks had helped prevent earlier crises since 1871 from leading to war. Moreover, alliances did not dictate participation: despite being their ally, Italy chose not to join Germany and Austria, instead declaring neutrality on August 3. The United States, which was not allied to any of the combatants, also opted for neutrality, President Woodrow Wilson telling the Senate on August 18, 1914, that the country "must be neutral in fact as well as in name." He had ignored pressure from Theodore Roosevelt, a former president, to go to war in response to the German violation of Belgian neutrality. In Europe, the Scandinavian states, the Netherlands, Portugal, Spain, Switzerland, Greece, Romania, and Bulgaria all declared neutrality. Reasons varied. Some powers, such as Italy, Romania, and Bulgaria, had territorial goals, but did not think the situation sufficiently clear and propitious to encourage them yet to take sides in order to pursue them.

In 1914, the key element leading to war was that Austria and Germany chose to fight, and Russia to respond. All three were empires with their constitutionalism held in check by practices of imperial direction, the latter providing characteristics that were autocratic and that ensured that small coteries of decision makers had great influence. As with Japan, these states were happy to see war with other major powers as a tool of policy, rather than as a means of distant overseas colonial expansionism that had only limited consequences for their societies, the position of the British, French, and American elites.[12] Foreign offices and diplomats played only a secondary part in the policies of the autocracies, as the prospect of war led military considerations to come to the fore, while the rulers and their advisers took the key role in arbitrating between contrasting attitudes and policies.

The aggressive and ambitious views of Wilhelm II were certainly important to the serious deterioration in Anglo-German relations.[13] There was a clear contrast between the commitment of the British elite, including the monarchs, Edward VII (r. 1901–10) and George V (r. 1910–36), to parliamentary democracy, and Wilhelm's antipathy to liberalism and parliamentary government, both of which were seen as weak. Edward's statue in Hobart, Tasmania, carries with it the inscription "The Peacemaker," which his personal diplomacy in furthering Anglo-French relations justified. The right-wing nationalists who looked to Wilhelm also despised parliamentary limitations, while Wilhelm's authoritarian position in the governmental system ensured that the extent to which such right-wing views were not, in fact, held by the bulk of the public could not determine policy. The moves to war were not taken by the elected representatives of the German populace.

Similarly, in 1915, the demand by opposition leaders in Bulgaria that the *Sŭbranie* (Assembly) be summoned before any decision for war be taken was ignored by the king; Aleksandar Stambolĭski, the agrarian leader, who called on the people and the army not to act, was imprisoned. In contrast, in 1914,

Carol I of Romania, a German by birth and marriage, wished to fulfill treaty commitments and support Germany and Austria, but the backing of most political leaders for neutrality led him to accept the constraints of his position as a constitutional monarch, and Romania initially stayed out of the war.

For the Germans, Russian mobilization increased the imminent danger of war and provided an opportunity to advance their interests by demanding its cancellation, a step that would have identified Russia as an inadequate ally, and thus have wrecked the Franco-Russian alliance.[14] When this demand was refused, war on Russia was declared on August 1. Russia's ally, France, then became the key element for the Germans. They issued an ultimatum, one that France could not accept, that France declare neutrality and provide guarantees for this neutrality, steps that would have destroyed the alliance with Russia and made France appear a worthless ally for any other power. The guarantees included their forts at Toul, Verdun, and elsewhere, which would have left France highly vulnerable. These forts provided France with protection from German attack across their common frontier and also provided bases from which to anchor any French attack on German-held Lorraine. France's refusal to accept the ultimatum led the Germans to declare war on August 3.

This declaration did not exhaust the bellicosity of those early August days. Operational factors dictated strategy for the Germans: the heavily defended nature of the Franco-German frontier, and the German need for speedy advances if sequential victory was to be obtained, with France overcome before Russia was defeated, led to a decision to attack France via Belgium. The flat terrain of much of Belgium north of the River Meuse was more appropriate for a rapid German advance than the hillier terrain of eastern France, while the Franco-Belgian frontier was poorly fortified. Belgium, however, rejected a German ultimatum to provide passage. This step ensured that the Germans would launch a violent invasion, instead of mounting an occupation, and that entailed commitments and delays that helped derange the German plan.

Belgian neutrality was guaranteed under the 1839 Treaty of London by the major powers, including Britain as well as Prussia, now Germany. The Germans hoped that Britain would not respond to the invasion of Belgium and were inclined to discount or minimize the risk. Bethmann-Hollweg was surprised by the British decision to go to war over what he termed "a scrap of paper," while Britain, with its small army, which had done conspicuously badly in 1899–1900 in the early stages of the Second Boer War against the Afrikaners of South Africa, anyway was a minor concern to German army planners. This view proved very foolish. There was no concern about the possibility of American intervention, which, indeed, could have had no immediate military effect. "Necessity knows no law," declared Bethmann-Hollweg when speaking in the *Reichstag* on August 4 about the invasion of

Belgium. "Necessity" had ensured that the German army had equipped itself with heavy howitzers and mortars before the war specifically to deal with the Belgian forts, especially around Liège.

The British government was far from keen on war, and, due to long-standing competing interests in Asia, there were serious tensions in Anglo-Russian relations despite an agreement of 1907 over spheres of influence in Persia.[15] Nevertheless, the government was unwilling to see France's position in the balance of power overthrown, and was concerned about the implications, if so, for Britain, while Germany's naval buildup had left the British profoundly distrustful of her expansionism. Indeed, this aggressive expansionism was regarded as more threatening than any particular calculations of the balance of power.

That concern did not necessarily mean war with Germany on behalf of France, which, in fact, was distrusted by many British strategists, as the debates in the Committee of Imperial Defence on a projected tunnel to France under the English Channel made clear.[16] However, British military planners had long been anxious that war between France and Germany would lead to a German invasion of the Low Countries that had to be stopped.[17] In 1914, it was unclear until late whether Britain would join the conflict, but the invasion of Belgium united most British political opinion behind the war. National honor was an important factor in the political culture and international realities of British politics.[18]

As in other states, different British political groups had particular views, with the governing Liberal cabinet focused on the issue of Belgian neutrality, as was public opinion, while the leadership of the Conservative opposition was particularly concerned about maintaining the Anglo-French entente. Some Conservative pressure groups, such as the National Service League, were more clearly anti-German.[19] On August 4, an ultimatum demanding the German evacuation of Belgium was issued. Prefiguring the outbreak of war with Germany in 1939, it was unanswered and led to British entry into the war, which thereby became a more wide-ranging conflict, and one that was to be far longer than the Germans had envisaged.[20]

The flawed expectation of how Britain would react, and the longer-term mishandling of British sensitivities anticipated the total misjudgment of Britain in both 1939 and 1940, and were aspects of the wider German failure to address the political aspects of any conflict. As the Germans sought, planned in great detail for, and anticipated a swift and decisive victory, in order to avoid the military, political, economic, and social complexities of a large-scale and lengthy war between peoples, the political dimension was not significant for their military planners, who, in any case, seriously underestimated their opponents' power, resolve, and behavior. An absence of rational assessment in the context of much wishful thinking, as well as calculations that in fact suggested to some that the planned short war was improbable, resulted, in Germany, in a countervailing attempt to control anxiety as well as risk, an attempt that led to a focus on planning that deteriorated into dogma and failure to note wider strategic and political parameters. [21]

Across Europe, cultural factors helped support the willingness of governments to declare war. Fatalism encouraged the resort to conflict, as did the potent cult of honor of the period, [22] and a militarism seen even, in its widest senses, in Britain, which did not have conscription, but which was a proud and confident nation at that time. [23] In Britain and elsewhere, national pride led to enthusiasm for war in 1914 alongside a powerful sense of duty to fight.

The likely nature of a major war was one that had long attracted commentary, with correspondence in the *Times* about the industrialization of warfare as early as 1870. Initially, a small number of European thinkers had anticipated the horrific casualties that developments in military methods and the expansion of army size were likely to produce. Frederick Engels had argued that the American Civil War (1861–65) indicated the likely destructiveness of future conflict between European powers, and he thought that this destructiveness would undermine existing state and class hegemonies and make revolution possible. In his *War of the Future in its Technical, Economic and Political Aspects* (1897), part of which was published in English as *Is War Now Impossible?* (1899), the Polish financier Ivan Bloch suggested that the combination of modern military technology and industrial strength had made great power warfare too destructive to be feasible, and that, if it occurred, it would resemble a great siege and would be won when one of the combatants succumbed to famine and revolution. Bloch argued that the stalemate on the battlefield that came from defensive firepower would translate into collapse on the home front.

The elder Moltke, the chief of the Prussian General Staff in the Wars of German Unification (1864–71), himself had become increasingly skeptical about the potential of the strategic offensive after 1871, and, presciently, was fearful that any major war would be a long one. [24] However, concerns about the consequent impact on casualty figures and military morale, and emphasis on the dangers of battlefield stalemate and of breakdown on the home front, only encouraged a focus on preventing the stalemate by winning the initial offensive and thus ensuring a short war. Indeed, in 1914, the dominance of thinking about the attack, rather than anything else, persuaded armies that the war could be won quickly.

That view suggests a degree of folly that was apparently to be underlined by the subsequent conduct of the war, but, in practice, military planners were well aware of the possibilities of defensive firepower and some of the remedies that were to be employed during the conflict were already in evidence. For example, German planners emphasized infantry-artillery coordination in the attack, as well as minimizing exposure to artillery fire by advancing in dispersed formations that coalesced for a final assault. Furthermore, it was observed that defensive strength could be challenged by field artillery operating in support of the attacking force.

However, the training of German reserves was much less developed than that of the regular units, and this problem helped explain the failures of 1914, for German war planning counted on the reserves to do much more than they were trained to do. There was a tendency to believe that the "German soldier" could and would do whatever was planned. Moreover, a major problem with the German military maneuvers and war games was posed by the need to satisfy the expectations of Wilhelm II, so that the defensive effect of machine-guns, for example, was often underplayed while the effect of the assault was overplayed. The maneuvers were contrived to show the strength of cavalry against infantry or machine-guns. On land and at sea, however, all powers sought to integrate new weapons, as well as improved weaponry, such as new artillery, and organizational means and systems, into their militaries, their maneuvers, and their plans, as part of the process by which they responded to advances and to apparent deficiencies. At the same time, a misguided confidence in cavalry continued.

The major war in the decade prior to the Great War was that between Russia and Japan in 1904–5, and it was followed carefully by foreign observers. They saw the war as a triumph for Europeanization in the form of Western military organization: the Japanese, whose army was modeled on the German and navy on the British, won by employing European military systems and technology more effectively than the Russians. Yet, the Japanese victory also came as a shock, in part because of Western racialist assumptions. Advocates of the offensive argued that the Russians stood on the defensive in Manchuria and lost, while the Japanese took the initiative, launched frontal assaults on entrenched forces strengthened by machine-guns and quick-firing artillery, and prevailed, despite horrific casualties.

Most commentators focused on tactical and operational factors, and overlooked the strategic dimension, notably the extent to which the land battles had not been decisive but only caused the Russians to fall back in Manchuria, while the Japanese, like the Germans in France in the winter of 1870–71, had been put under great pressure by the continuation of the war, which they could not afford.[25] Japanese victory in practice owed much to political weakness in St. Petersburg in 1905, notably a revolution there, in part fostered by Japanese military intelligence, rather as German sponsoring of the Bolsheviks played a role in 1917. However, this strategic dimension behind Japanese victory was neglected, while Japanese success on the battlefield and in the war ensured that the tactical superiority of the defense was further underplayed. Given contemporary racist attitudes, European experts concluded that the infantry of the superior races of Europe would be capable of at least similar deeds, albeit at heavy cost, maybe a third of the army.[26]

European army leaders assumed that troop morale and discipline would enable them to bear such losses, but the expectation of them provided the impetus for programs to expand army size, including discussion in Britain about the need for conscription or more volunteering. The expansion in army size across much of Europe took forward the example created by the German army in the Wars of Unification (1864–71), and also drew on the major growth in the population of the West and Japan. The military, however, failed to anticipate how long a major conflict might be—or anticipated it but omitted to tell civilian politicians for fear that, if they knew, they would never contemplate war as an option.

World War I was intended by each of its participants as a short and manageable, albeit costly, international war between regular forces, indeed as a reprise of the Franco-Prussian and Russo-Japanese wars. It was understood that there would be heavy casualties due to the nature of military technology, a point driven home by the Russo-Japanese War, but it was believed that a speedy victory could be delivered by the side that attacked. Speedy victory, the ideas of Nietzsche calling on what were understood as those of Clausewitz in the case of Germany, was what was attempted by all the major powers in 1914 once war had begun.

That did not mean that all the powers were equally to blame for the war. The anniversary in 2014 of the opening of the war led to even more division among scholars than might have been anticipated. This division related not only to discussion about the nature of the conflict, but also to the responsibility for the war. This debate, often contentious in character, was not always terribly helpful. Indeed, much of it repeated themes already seen in the 1920s in the bitter discussion over war guilt.

There was also a widespread failure adequately to integrate into the discussion of the diplomacy the often-excellent scholarly work on the military preparations for the conflict. In particular, the nature of Austrian and German planning scarcely accorded with the unconvincing argument that Europe somehow slid or sleepwalked into war, an approach that turns perpetrators into victims and satisfies national myths of victimhood.[27] Instead, a degree of preparedness encourages support for the argument that war was intended. In Austria, the army leadership, which believed that war would be socially rejuvenating, failed to inform civilian ministers about the reality of the military situation.[28]

This approach suggests that the war was not so much a failure of statecraft, as some have argued, but, instead, a breakdown of deterrence accompanied with a willingness by key figures to turn to war. If these latter were combined with a strategic confusion at the heart of decision making, then the combination was deadly. Alongside this came the extent to which the dynamic of the crisis meant that constructive ambiguity was no longer credible. Moreover, it is unhelpful to assume that there were system-appropriate policies; this approach underplays the major policy divisions within decision-making elites.[29]

While planning for a war is not simple proof of intent, pressure from the military was highly significant. The decision makers had lost the sense of the fragility of peace and order. The politicians used the threat of war to put the other side (notably internationally but sometimes also domestically) at a disadvantage with the aim of increasing their own leverage, and, in doing so, they miscalculated. Everyone made mistakes. Austria wanted war (albeit only a Balkan one); Germany, living in a militaristic bubble, was criminally negligent and, partly through fumbling, created a situation in which she could and would not rein in Vienna or retreat in the face of Franco-Russian opposition, the Kaiser reproaching Bethmann-Hollweg on July 26 for the mess that had been created; France was rigid; and Russian policy was incompetent. Britain comes out better than others on balance. With respect to the British government, criticisms fail to consider simple parliamentary arithmetic. Any attempts to issue a warning to Germany before the invasion of Belgium that such an invasion would bring Britain into the war were likely to be hollow because of the makeup of the cabinet and because the governing Liberals were dependent on Labour and Irish support for a majority in the House of Commons. The inherent logic of Britain's geopolitical position, both European and global, meant that, once the fighting had started with a German attack in the west, she had to enter the war.

There was long a complaint among British military historians that there was a major disjuncture between their work, notably on the learning curve of the British army during World War I, and, on the other hand, the understanding of the war in contemporary popular culture. Now as an example of differing tramlines, we can add much of the recent work on the diplomatic background to the war. The military dimension had been largely sidelined in this work. Thus, Gordon Martel argued, "Premeditation is not to be proven by the existence of war plans or by the warlike pronouncements of military men. Strategists are expected to plan for the next war: the politicians and diplomats decide when that war is most likely to occur."[30] However, in his work and elsewhere, in what was very much diplomatic history of a traditional type (not itself a cause of criticism), there was a tendency to find the answer in the material that was studied. Moreover, this tendency led to a particular slant: that of individual and collective faults in 1914, the latter very much set in motion by the former. These faults thus lead to a degree of collective responsibility, doubtless a conclusion suitable for our "transnational times."

As far as Britain is concerned, there is blame for Sir Edward Grey, the foreign secretary, not least for failing to send clear messages, and this failure is presented as important in what was a collective malaise affecting all the European powers. In terms of the particular diplomatic processes that judgment may well appear reasonable, but is it sufficient? With respect to Grey, criticisms of him fail to consider simple parliamentary arithmetic. Moreover, German confidence that victory would have been won before British units could be deployed on the continent in any significant numbers, as well as a lack of concern about British naval moves and the fate of German colonies, would have lessened the impact of such a warning.

Military planning, procurement, and preparations in the situation in 1914, let alone military influences in the decision-making process and cultural bellicosity were present for all powers—even the Swiss mobilized; but they were crucially different in character, context, and consequences. Moreover, this difference can be underplayed. Moltke considered using the opportunity for war but did not push for it until late July. What mattered was the balance of forces at Berlin. For much of July, the civilians held the military in check. More profoundly, the obvious contrast in considering war-planning and the move to war is between France and Britain on one side, and Austria, Germany, and Russia on the other; with the difference linked to the nature of the individual states. It is instructive to note, for example, that prewar, the French government had decided not to pursue the military option of advancing against Germany via neutral Belgium, while Germany took a very different view of conflict with France and made its attacking military operational plan central to its war strategy. Yet, that point about political system is insufficient for, even in the case of Austria, Germany, and Russia, there were important contrasts. In particular, what mobilization meant for Germany was very different to what it entailed for Russia, a point widely neglected in much of the discussion in 2014. For the former, mobilization was a move to immediate conflict that was not the same as for Russia.[31]

Such distinctions are important because they counter a widespread intellectual tendency, considering both past and present, to focus on the supposed faults of "the system," rather than of particular actors and groups within it. The consequence is a form of transferred responsibility, so that, to take another prominent example, Appeasement by Britain and France in the 1930s is somehow made responsible for Hitler's expansionism and for Stalin's decision to join in himself in this expansionism. This approach takes responsibility from where it truly lies, with Hitler and Stalin.

In 1914, the British sought to rely on the traditional means of addressing an international crisis, that of the Concert of Europe seen from the seventeenth century and, more particularly, in the nineteenth. This approach indeed had succeeded in the case of the First Balkan War (1912–13) in preventing a wider war. However, in 1914, operating to different purposes and on other timetables, Austria and Germany were unwilling to follow this course. Their policies and attitudes caused the war, and not the varied errors of the statesmen struggling with the developing crisis. It is also of course necessary to locate this German preference in the political and cultural bellicosity that was so strong in Germany in particular in the early 1910s. A fervent national patriotism, a belief that war was a natural state, and a conviction that it was honorable to die for one's country were linked to a strong fear of falling behind and a sense that the opportunity that existed to attack might not continue.

British participation in the war helped ensure that Germany would not win. This participation was not itself responsible for the serious failures to achieve victory in 1914 at the operational level, notably that of Germany, but was of strategic significance from the outset. That the attempt for victory, nevertheless, was continued, with a degree of success against weaker states, until finally it led to the defeat of all the German-led Central Powers in 1918 is a dimension that deserves attention. There is an obvious contrast in duration with the Franco-Prussian (1870–71) and Russo-Japanese (1904–5) wars, but this was due not to a change in weapons technology but to the scale of conflict. The key scale was strategic, not operational, notably the extent to which alliance systems made it difficult to isolate a conflict and, instead, ensured that a military-political outcome involved the overthrow of an entire alliance. But for that it was possible that the 1914 German campaign would have led to the overthrow of France, rather as the 1870 one had done. The involvement of an alliance made World War I, and, indeed, World War II, very different to any major war since the Napoleonic Wars; and in the latter the alliance system had essentially been one-sided in that it was the case of Napoleon's opponents.

The primacy of the political dimension of strategy was demonstrated by the failure of the German alternative of dictating victory by means of pushing military factors to the fore. Germany subordinated political to military considerations in bringing Britain and the United States into the war against her respectively in 1914 and 1917. In each case, although policies were justified on military grounds, those of advancing more easily via Belgium and of trying to knock out British trade by submarine attack respectively, this was a serious strategic mistake and one that contributed greatly to eventual German failure.

The German campaign had failed already in 1914 before the Allied counterattack in the battle of the Marne and the subsequent stabilization of what became a Western Front in France and Belgium. This failure was because Britain's entry into the war promised that what was already, due to Russian involvement, a two-front war would become a longer and more difficult struggle. This was so as long as France did not collapse, as it was to do in 1940 when Germany only faced a one-front war. In 1864–71, there had been no two-front wars for Prussia/Germany. Moreover, British involvement meant that German resources would be diverted to the navy.

By late 1914, the carefully prepared German prewar strategic planning appeared precarious and over-optimistic. A dangerous overconfidence was apparent at every level. The Germans were apt to consider the enemy a "constant" instead of an "opposing variable" that was able and willing to respond rapidly, as the French eventually did by moving troops from their right flank to help defend Paris. Germany had wrongly envisaged a repeat of the Franco-Prussian War, with France collapsing, having suffered similar command failures to those in 1870. Austria similarly hoped that a war would be quick and decisive. As should have been anticipated, nothing really played out according to the script; the opponents of Austria and Germany proved to be tougher to break than anticipated, which led to exhaustion and very high casualties. Austria failed badly in an attempt to conquer Serbia, whose army was less well-resourced but much better-commanded,[32] and Austria was also put under heavy pressure by Russia.

SUBSEQUENT ENTRIES

With subsequent entries into the war in Europe, nationalism and power politics interacted in the prism of individual and group hopes and fears.[33] Control over territory was a key bargaining basis. In 1914, Italy had not come to the aid of Germany and Austria, its allies since 1882. Instead, it was won over by the Treaty of London, signed on April 26, 1915, by which Britain, France, and Russia promised Italy extensive gains from Austria: the Trentino, South Tyrol, Trieste, Gorizia, Istria, and northern Dalmatia. This meant that those territories, presented in Italy to the public as the last stage of the war for independence from Austria, had to be conquered. Germany had bullied Austria into offering the Trentino to Italy, but Austria was not willing to match the Allied offer elsewhere.

Only the Socialist Party opposed the war when it was voted on by the Italian Parliament on May 20. The remainder of the political world wished to see Italy become a great power, and Antonio Salandra, the conservative prime minister, presented Italy's policy as "sacred selfishness." Benito Mussolini, the editor of the Socialist Party newspaper *Avanti*, was expelled from the party because of his support for the war, which he saw as a reconciliation of patriotism and socialism. Instead, with support from Italian and French industrialists, Mussolini launched the interventionist paper *Il Populo d'Italia*. Italy declared war on Austria on May 23 and on Turkey on August 21; although not on Germany until August 27, 1916; the government did not wish to provoke the Germans to send more troops to the Italian front.

Bulgaria, in turn, joined the German alliance in 1915. This was part of the equations of territorial offers. The entry was related to the position of Romania. Bulgaria's entry into the war had been pressed for by the Hungarian political establishment, which had resisted support from Wilhelm II and Franz Josef for the Romanian offer of continued neutrality in return for autonomy for Transylvania. It was the part of the Habsburg empire with a Romanian majority, a part within the kingdom of Hungary. Instead, the Hungarian leadership argued that Bulgarian entry would isolate and neutralize Romania. The Bulgarians were promised Serbian Macedonian and Serbian territory on the left bank of the River Morava by Germany and Austria, who also successfully pressed Turkey to offer Bulgaria territory. In contrast, Russia had supported Serbia in opposing promised gains for Bulgaria in Macedonia if it joined the Allies.

The entry of Bulgaria and Italy into the war did not reflect a commitment to particular alliances, but, rather, the continued determination and perceived need for (very different) second-rank powers to make assessments of opportunity. Far from the perceived ideology of either alliance playing a role, the key element was the possibility for the gain of territories. They were small in themselves but were made important as a result of nationalist public myths. Their gain was seen as a sign of national success, one that justified the regime.

In turn, Romania entered the war in 1916. Its government sought Transylvania and Bukovina from Austria and was encouraged to enter the war by Russian success that year. By going to war, the Romanians hoped to benefit as a victor in what might be an imminent peace. On August 27, Romania declared war on Austria, but Germany declared war on Romania the following day, and was followed rapidly by Bulgaria and Turkey. Most of Romania was swiftly overrun.

Diplomatic and military pressure could be important. Under British pressure, Portugal entered the war on the Allied side in 1916. In practice, this owed much to German action. A British request had led Portugal to intern German and Austrian ships in Lisbon on February 23. In response, Germany declared war on March 9, and Austro-Hungary on March 15. In Greece in 1917, King Constantine I was deposed with Allied support in June and the country, under his rival, the pro-Allied British Prime Minister Eleftherios Venizelos, declared war on Germany.

AMERICA ENTERS THE WAR, 1917

American hostility was more serious. Having failed to drive France from the war at Verdun in 1916 and experienced the lengthy and damaging British attack in the Somme offensive the same year, the Germans sought to force Britain from the war by resuming attempts to destroy its supply system. There was a parallel with the invasion of France via Belgium in 1914, in that the strong risk that a major power would enter the war as a result, Britain in 1914 and America in 1917, was disregarded on the grounds that success could be obtained as a result of the German attack such that the war could be ended before the new enemy became an acute danger. In 1917, however, the Germans, unlike in 1914, had had plentiful warnings as a result of their earlier use of unrestricted submarine warfare in 1915 and the impact on American policy. There was also a failure of planning; anticipated outcomes from the submarine assault did not arise, and the timetables of success miscarried.

Yet, this account assumes a rationalist balance of risks and opportunities that ignores the extent to which the decision to turn to unrestricted submarine warfare reflected an ideology of total war and a powerful Anglophobia based on nationalist right-wing circles, for example the Pan German League, which saw British liberalism and capitalism as a threat to German culture. These ideas were given political bite by the argument that the German government, notably the chancellor, Bethmann-Hollweg, was defeatist and interested in a compromise peace, which he indeed hoped for in late 1916 via American mediation, and that support for unrestricted warfare was a sign of, and security for, nationalist commitment, a view later to be taken by the Nazis when they pressed for total war.

Grand Admiral Alfred von Tirpitz, who had links with nationalist right-wing circles, was allowed to resign as secretary of state of the Imperial Naval Office in March 1916 because his support for unrestricted submarine warfare had led him to quarrel with the chancellor and to challenge the position of the kaiser.[34] The following month, an American protest resulted in Wilhelm II ordering the suspension of the permission given that February for the sinking without warning of armed freighters—but not of passenger ships.

On January 31, 1917, however, despite Bethmann-Hollweg's opposition, Germany announced, and on February 2 resumed, unconditional submarine warfare, which led to America breaking diplomatic links the next day and declaring war on Germany (but not its allies) on April 6. Congress had approved the decision, although six senators and fifty congressmen opposed it. The German military leadership, increasingly politically influential, was unsympathetic to American moralizing, while, as in 1941, there was also the view that America was already helping the British and French war effort as much as it could commercially, which was the case to a considerable extent. Moreover, there was a conviction that Britain could be driven out of the war rapidly by heavy sinkings of merchantmen: a belief that the submarines could achieve much, and that this achievement would have an obvious consequence. It was claimed that the British would sue for peace on August 1, 1917. Furthermore, many German submarine enthusiasts assumed that their force would be able to impede the movement of American troops to Europe very seriously and, more generally, there was a failure to appreciate American strength.[35] Thus, Germany in 1917 prefigured the situation of Germany and Japan in 1941, and with similar dire consequences.

In 1914, there was active hostility in America to the idea of participation in the European war. It was seen as alien to American interests and antipathetic to her ideology, although the liberal credentials of American policy were rather tarnished by interventions in Mexico in 1914 and 1916, Haiti in 1915, and the Dominican Republic in 1916. These interventions reflected and camouflaged imperialist assumptions, but not a drive for territorial expansion. The unrestricted submarine warfare that sank American ships (and also violated international law) led to a major shift in attitudes in which Americans became persuaded of the dangerous consequences of German strength and ambitions and did so in a highly moralized form that encouraged large-scale commitment. Thus, the United States constructed national interest in terms of the freedom of international trade from unrestricted submarine warfare.

Germany's crass wartime diplomacy exacerbated the situation, notably an apparent willingness to divert American strength by encouraging Mexican opposition, including *revanche* for the major losses suffered in the Mexican-American War of 1846–48. The Americans were made aware of this when the British intercepted a telegram to the German ambassador in Mexico from Arthur Zimmermann, the foreign minister. The logic of this apparently bizarre move was preemptive: the Germans wished to distract American energies from war in Europe. Yet again, strategic factors were ignored in favor of operational considerations. American sensitivity about German links with unstable Mexico was acute: later that year, the *San Francisco Chronicle* claimed (incorrectly) that German submarines were being constructed at a secret base in Mexico.

The United States had given neutrality added legitimacy to other states.[36] In turn, after she broke off diplomatic relations with Germany in February, she invited all neutral countries to do the same, and this appeal had some success. Aside from Latin American states breaking off diplomatic relations with Germany, others followed the United States in declaring war, including Cuba and Panama, both American client states, on August 7, and Brazil, which also suffered from the unrestricted submarine warfare, on October 26, 1917. However, the Brazilian contribution was more modest than in World War II when about 25,000 troops were sent to fight. In World War I, in contrast, only a small Brazilian naval squadron was eventually dispatched, and it did not see active service.[37] Nevertheless, the Brazilian declaration of war contributed to Allied commercial warfare against Germany. Guatemala, Nicaragua, Costa Rica, Honduras, and Haiti followed in declaring war in 1918. The United States had declared war on December 7, 1917, on Austria, but did not follow suit against Bulgaria or Turkey.

Seeking an opportunity to enter the conflict, China broke off diplomatic relations with Germany on March 14, 1917, and declared war on Germany and Austria on August 14; Siam (Thailand) had done so on July 22. Encouraged by Marshal Foch, chief of the French General Staff from May 1917, the French proved receptive to the Chinese idea of an expeditionary force, but, although laborers in the "Chinese Labor Corps," active since 1916, were acceptable, the other Allies were opposed to the dispatch of Chinese troops, for political reasons and because of concerns about fighting quality and the transportation burden, and the plan failed.[38] Japan was determined to retain its political position in China and was unwilling to see the latter contribute to the war effort. The following year, Ethiopia offered to declare war on the Central Powers in return for a role in the peace talks and new weaponry, but this offer, which challenged Italian ambitions for control over the country, was rejected.

Other powers remained neutral in 1917, finding that the pressures of what was seen as the total clash of industrialized societies pressed hard on their trade, not least as a consequence of the Allied blockade of Germany and of German submarine warfare. The Netherlands and Denmark, both of which were maritime trading powers bordering Germany, and thus key challenges to the Allied blockade of Germany, found themselves under contrary pressures. In particular, their use of diplomacy had to be accompanied by careful consideration of the military situation, for example the possibility of German invasion, and also of economic interests.[39]

AFTER WORLD WAR I

The causes of World War I, the "Great War," dominate attention for this period, but there is also the need to address the conflicts that followed the war. Most of these arose from its consequences, both in terms of the Russian Revolution of 1917 and with regard to the peace settlement, which is generally referred to as the Peace of Versailles. Each was disputed, and some of the conflicts were large-scale, notably the Russian Civil War, the Russian invasion of Poland,[40] and the Greco-Turkish war. Fourteen foreign powers sent troops to help the anti-Communist side in Russia, but only limited resources were committed, not least thanks to the general unpopularity at home of intervention.

A separate but related series of conflicts, stretching from Morocco to Afghanistan, reflected the unease across the Islamic world at the extension of Western power that in part arose from the peace settlement, but that also had separate, specific causes, as in both resistance to the extension of Spanish control in northern Morocco, and the Third Anglo-Afghan War.

As also with conflict in both Mexico and China in the late 1910s and 1920s, war should be seen not as the breakdown of systems of peace and practices of deterrence, but rather as a product of the willingness, indeed eagerness, to fight, and the additional encouragements offered by the large numbers of men habituated to fighting by the recent war and the plentiful supplies of armaments. Ideology was very much to the fore in the conflicts focused on Communism, which included the Romanian overthrow of a Communist regime in Hungary in 1919, as well as those relating to the Islamic rejection of Western control as in Iraq in 1919 and Syria in 1925–26. Ideology was far less present in the struggles between Chinese generals (and their Mexican counterparts) over whom would dominate government. Indeed, those struggles, which in the case of China were civil wars in the 1920s on a major scale, were a form of politics in the absence of the legitimation provided by governmental systems that had been overthrown in the early 1910s.

The explicitly anti-Christian nature of Russian Communism pushed religion to the fore in the reaction to it, as in Poland. Indeed, it would not be fanciful to see a new period of religious warfare as breaking out in the late 1910s. The key characteristic of ideological conflict that was readily apparent was that of the interrelationship of international and domestic warfare. This was very much the Communist thesis and rested on a concept of continuous class conflict directed by a higher power (the Politburo) seeking to implement an all-encompassing plan. In that approach, the "system" came first, a system of never-ending conflict. This was the "Cold War," which, in its opening stages, was far from limited or "cold."

NOTES

1. Michael Paris, *Warrior Nation: The Representation of War in British Popular Culture, 1850–2000* (London: Reaktion, 2000).

2. Christopher Forth, *The Dreyfus Affair and the Crisis of French Manhood* (Baltimore, MD: Johns Hopkins University Press, 2004).

3. Azar Gat, *A History of Military Thought from the Enlightenment to the Cold War* (Oxford: Oxford University Press, 2001), 343.

4. Paul Schroeder, "A. J. P. Taylor's International System," *International History Review* 23 (2001): 25.

5. John Röhl, *Wilhelm II: Into the Abyss of War and Exile, 1900–1941* (Cambridge: Cambridge University Press, 2014).

6. Michael Howard, *The Franco-Prussian War: The German Invasion of France, 1870–1871* (London: Methuen, 1961), 220–21.

7. D. N. Collins, "The Franco-Russian Alliance and Russia's Railways, 1891–1914," *Historical Journal* 16 (1973): 777–88.

8. Matthew Seligmann, "A View from Berlin: Colonel Frederick Trench and the Development of British Perceptions of German Aggressive Intent, 1906–1910," *Journal of Strategic Studies* 23 (2000) and *Spies in Uniform: British Military and Naval Intelligence on the Eve of the First World War* (Oxford: Oxford University Press, 2006).

9. Anika Mombauer, *Helmuth von Moltke and the Origins of the First World War* (Cambridge: Cambridge University Press, 2001).

10. Thomas Otte, "'What We Desire Is Confidence': The Search for an Anglo-German Naval Agreement, 1909–1912," in *Arms and Disarmament in Diplomacy*, ed. Keith Hamilton and E. Johnson (London: Valentine Mitchell, 2007), 47; Matthew Seligmann, ed., *Naval Intelligence from Germany: The Reports of the British Naval Attachés in Berlin, 1906–1914* (Aldershot: Routledge, 2007).

11. W. D. Godsey, "Officers vs Diplomats: Bureaucracy and Foreign Policy in Austria-Hungary, 1906–1914," *Mitteilungen des Österreichischen Staatsarchiv* 46 (1998): 43–66.

12. Richard Hamilton and Holger Herwig, ed., *Decisions for War, 1914–1917* (Cambridge: Cambridge University Press, 2004); William Mulligan, *The Origins of the First World War* (Cambridge: Cambridge University Press, 2010).

13. Roderick McLean, *Royalty and Diplomacy, 1890–1914* (Cambridge: Cambridge University Press, 2001).

14. Jack Levy and William Mulligan, "Shifting Power, Preventive Logic, and the Response of the Target: Germany, Russia, and the First World War," *Journal of Strategic Studies* 40 (2017): 731–69.

15. Keith Neilson, *Britain and the Last Tsar: British Policy and Russia, 1894–1917* (Oxford: Oxford University Press, 1996).

16. Keith Wilson, "The Channel Tunnel Question at the Committee of Imperial Defence, 1906–1914," *Journal of Strategic Studies* 13 (1990).

17. Thomas Otte, "'The Method in Which We Were Schooled by Experience': British Strategy and a Continental Commitment before 1914," in *The British Way in Warfare: Power and the International System, 1856–1956*, ed. Keith Neilson and Greg Kennedy (Farnham: Ashgate, 2010), 318–19.

18. Thomas Otte, "Neo-Revisionism or the Emperor's New Clothes: Some Reflections on Niall Ferguson on the Origins of the First World War," *Diplomacy and Statecraft* 11 (2000): 285.

19. Frank McDonough, *The Conservative Party and Anglo-German Relations, 1905–1914* (Basingstoke: Palgrave, 2007).

20. Keith Wilson, ed., *Decisions for War, 1914* (London: UCL Press, 1995).

21. Dennis Showalter, "From Deterrence to Doomsday Machine: The German Way of War, 1890–1914," *Journal of Military History* 64 (2000): 708.

22. Avner Offer, "Going to War in 1914: A Matter of Honour?" *Politics and Society* 23 (1995): 213–41.

23. R. H. Macdonald, *Sons of the Empire: The Frontier and the Boy Scout Movement, 1890–1918* (Toronto: Toronto University Press, 1993); Graham Dawson, *Soldier Heroes: British Adventure, Empire and the Imagining of Masculinities* (London: Routledge, 1994).

24. Antulio Echevarria, *Imaging Future War: The West's Technological Revolution and Visions of War to Come, 1880–1914* (Westport, CT: Greenwood, 2007).

25. Gary Cox, "Of Aphorisms, Lessons, and Paradigms: Comparing the British and German Official Histories of the Russo-Japanese Wars," *Journal of Military History* 66 (1992): 389–401.

26. BL Add. 50344, p. 3; Michael Howard, "Men against Fire: The Doctrine of the Offensive in 1914," in *Makers of Modern Strategy from Machiavelli to the Nuclear Age*, ed. Peter Paret et al. (Princeton, NJ: Princeton University Press, 1986), 510–26.

27. Christopher Clark, *The Sleepwalkers: How Europe Went to War in 1914* (London: Allen Lane, 2012).

28. Lawrence Sondhaus, *Franz Conrad von Hötzendorf: Architect of the Apocalypse* (Boston, MA: Humanities Press, 2000); John Zametica, *Folly and Malice: the Habsburg Empire, the Balkans and the start of World War One* (London: Shepheard-Walwyn, 2017).

29. Thomas Otte, *July Crisis: The World's Descent into War, Summer 1914* (Cambridge: Cambridge University Press, 2014). For political science approaches, Jack Levy and John Vasquez, eds., *The Outbreak of the First World War: Structure, Politics, and Decision-Making* (Cambridge: Cambridge University Press, 2014); John Vasquez, *Contagion and War: Lessons from the First World War* (Cambridge: Cambridge University Press, 2018) and "The First World War and International Relations Theory: A Review of Books on the 100th Anniversary," *International Studies Review* 16 (2014): 623–44.

30. Gordon Martel, *The Month That Changed the World, July 1914* (Oxford: Oxford University Press, 2014), 428.

31. Dominic Lieven, *Towards the Flame: Empire, War and the End of Tsarist Russia* (London: Allen Lane, 2015).

32. James Lyon, *Serbia and the Balkan Front, 1914: The Outbreak of the Great War* (London: Bloomsbury, 2015).

33. "Forum on the Spread of War, 1914–1917: A Dialogue between Political Scientists and Historians," *Foreign Policy Analysis* 7 (2011): 139–216.

34. Matthew Stibbe, *German Anglophobia and the Great War, 1914–1918* (Cambridge: Cambridge University Press, 2001); Raffael Scheck, *Alfred von Tirpitz and German Right-Wing Politics, 1914–1930* (Atlantic Highlands, NJ: Brill, 1998).

35. Lawrence Sondhaus, *German Submarine Warfare in World War I: The Onset of Total War at Sea* (Lanham, MD: Rowman & Littlefield, 2017).

36. Roger MacGinty, "War Cause and Peace Aim? Small States and the First World War," *European History Quarterly* 27 (1997): 46–47.

37. Stefan Rinke, *Latin America and the First World War* (Cambridge: Cambridge University Press, 2017).

38. X. Guoqi, "The Great War and China's Military Expedition Plan," *Journal of Military History* 72 (2008): 124–38.

39. Maartje Abenhuis, *The Art of Staying Neutral: The Netherlands in the First World War, 1914–1918* (Amsterdam: Amsterdam University Press, 2006).

40. Jochem Boehler, *Civil War in Central Europe, 1918–1921: The Reconstruction of Poland* (Oxford: Oxford University Press, 2018).

Chapter Seven

World War II and Its Origins, 1931–45

BACKGROUND

Debate over the causes of World War II links contemporaries with those born after 1945. For contemporaries, such debate was largely political, an attempt to mobilize support, both domestic and international, behind the war effort. For subsequent generations, in contrast, debate links the issue of war guilt, on the part of wartime opponents and others, to more general questions of justification and vindication and to related political dynamics. Consideration of the war certainly demonstrates that the combatants, and their alignments, were far from inevitable and were definitely not seen in that light by contemporaries. Therefore, the discussion of how these alignments arose is a key issue in the politics involved in the war and its recollection. This discussion also relates to postwar debates over responsibility.

Wartime alignments are crucial to the process by which guilt or praise are apparently established by association, with motivation a key way in which causation is discussed. For example, Hungary appears "bad" because it allied with Germany against the Soviet Union in 1941, whereas the Soviet Union is apparently vindicated for posterity because, although it allied with Germany in 1939, invading Poland, it was attacked by Germany in 1941. In practice, an understanding of the policies of these and other states requires a subtler consideration of their situation, politics, and options. Eastern European powers had their own history and interests to consider, and thus their own causes of war. Hungary aligned with Germany in order to regain territories lost, in the Treaty of Trianon in 1920, as a consequence of being on the losing side in World War I.

Constraints and/or opportunities, and their consequences, can be emphasized in the assessment of the Eastern European powers. It is certainly necessary to note that there was an element of choice. Thus, the politics of Romania can be criticized not simply because of alliance with Germany against the Soviet Union from 1941 to 1944, but also due to the policies it followed, notably genocidal anti-Semitism. Tens of thousands of Jews were slaughtered when the invading Romanians captured the city of Odessa in 1941.

Debate over the causes of World War II is a particularly difficult subject because of the number of different conflicts involved. This number is reflected in the widely contrasting titles, dates, and periodization offered for the war, as well as the dangers of assuming clear causal links between these conflicts and of arguing that there was a common struggle. Whereas the British, French, Germans, and Poles date the war from 1939, the Chinese turn to 1937, when they were attacked by Japan, and there are Spanish commentators who see the Spanish Civil War (1936–39) as the first stage of World War II. Italy entered the war in 1940, but the Soviet Union and, later, the United States did not enter the war until 1941, while the Soviets did not declare war on Japan until the closing days of the war in August 1945. Furthermore, the debate over the causes has a political dimension because of the continuing significance of issues of responsibility and guilt. As a result, the work of historians is located, at least for the public, within continuing controversies about blame for aspects of the war.

JAPAN ATTACKS CHINA

The interactions of ideological inheritance, strategic culture, institutional drives, and political and economic circumstances were shown in Japanese bellicosity in the 1930s. Hit, in the Great Depression, by a collapse in vital exports and mass unemployment, Japan became more bellicose, breathing, as the British diplomat Sir Victor Wellesley put it in 1932, "an atmosphere of gun-grease."[1] In part, this bellicosity arose because sections of the military followed autonomous policies ignoring civilian restraint, and in part because the military as a whole supported a militarism that challenged civil society and affected government policies, leading to what was called, by some, "government by assassination."

In 1930–31, Japanese soldiers, angered by what they felt was the failure of the government to defend national interests, planned a coup and the creation of a military government. Instead, they found it easier to seize control of Manchuria, where they were concerned that Japan's existing leased rights were under threat. On September 18, 1931, Japanese soldiers of the Kwantung army blew up part of the South Manchurian railway near the city of Shenyang (Mukden). This incident served as the basis for taking over the city and for an advance north and west from the Kwantung Leased Area, the established Japanese sphere of influence in southern Manchuria. The Japanese separately increased pressure on China by attacking the Chinese section of Shanghai in 1932.

A lack of support by the government in Tokyo led to the assassination of the prime minister in 1932, and civilian views became less important in the government. Party politics continued but became less significant, and in 1936 the assassination of officers in an attempted coup by younger officers served to quiet voices of reason and caution among both civilians and military. Moreover, the impact of the Depression, and the popular relief at what was seen as decisive action in Manchuria, both hit Japanese internationalism. The League of Nations criticized Japanese policy, which led to Japan leaving the League in 1933 in its most public abandonment of internationalism, but suffering no effective consequences.[2]

Success in Manchuria did not suffice to guarantee Japan's security and satisfy its ambition. In 1933, the neighboring province of Jehol (modern Hebei) was invaded. A conviction that large-scale war between Japan and the United States or Soviet Union was inevitable had developed in Japanese military circles and led to pressure for the strengthening of the military, the state, and Japanese society, with the latter two seen in an authoritarian and militaristic perspective. China was regarded as a base for the vital resources necessary for preparing for this conflict. Manchuria was developed by the Japanese army as a military and industrial base that was outside civilian control and used as a basis for expansionism in China and for greater strength in the event of possible war with the Soviet Union, which was greatly feared in Japanese army circles. A military-industrial complex in Manchuria, drawing on its coal and iron, served as the basis for a system of planning that sought to encompass industrial capacity.[3]

There was also concern in Japan about military developments in China, notably the expansion of Guomindang power and pretensions, especially the buildup of the German-trained divisions of the Central Army. In practice, Republic of China leader Jiang Jieshi (Chiang Kai-shek) proved willing to accommodate Japanese demands in the early 1930s, in large part because he wished to strengthen and unify China, not least by destroying the rival Communists. Jiang saw this as a necessary step before confronting Japan, but it turned out to be an intractable goal.

Some in the Japanese army were inclined to despise their opponents and to exaggerate both the significance of their "honor" and the ability of their will and military machine to overcome the problems posed by operating in China, an attitude that prefigured that of the Germans toward the Soviet Union in 1941. These problems were not simply a matter of the ratio between Chinese space and Japanese resources, important as that was, but also of the determination and fighting quality of the Chinese. These Japanese were misleadingly confident that China would fall rapidly, although others in the army knew better.

In the event, conquest turned out to be an impossible goal: the limited warfare and amphibious-based expeditions of Britain and France in the nineteenth century were a more realistic response to the nature of the military balance between China and the outside powers, and to the problems of campaigning in China. A similar point could be made about the Anglo-French Crimean War attacks in 1854–56, as opposed to attempts to conquer Russia itself in 1812, 1919–20, and 1941–42.

Overlapping with and helping to mold discussion in Japan of policy in "rational terms," there was both a social basis for militarism in what remained mainly an agrarian society,[4] and a sense of imperial mission, a sense that for some, but far from all, was linked to a radical Shintō ultranationalism, and, in particular, a belief in a divine providential purpose of Japanese superiority and expansion and a god-given purpose and right to rule.[5] The attitudes and impact of Zen Buddhism have been a source of controversy, but one school of scholarship emphasizes their bellicosity.[6]

After the violence of 1930–36, it is difficult to speak of moderates in the Japanese government. In particular, compromise was regarded as failure. However, the Sino-Japanese War that broke out in 1937 was unintended. An unplanned incident between Chinese and Japanese troops near the Marco Polo Bridge outside Beijing on July 7–8 during Japanese night exercises occurred at a time when many Japanese leaders favored preparing for war with the Soviet Union and Communism. Ideally, it was felt that, as part of this preparation, China should be persuaded to accept its fate as a junior partner of Japan. Jiang's uncooperativeness prompted Tokyo to try to give him a short sharp lesson.

Jiang's attitudes reflected a recent strengthening of Chinese nationalism, as well as the marked improvement in the German-trained Chinese army. He moved divisions north across the Yellow River into the part of China that in 1933 had been left demilitarized by the Chinese Central Army, and thus toward the Japanese. While not wanting large-scale conflict, the Japanese government felt the nation's honor had been challenged by the Marco Polo Bridge clash and sent fresh forces to the region. Refusing to yield to Japanese pressures to withdraw, Jiang proved willing to fight and eager to assert Chinese sovereignty, and large-scale conflict broke out in late July as the Japanese moved large numbers of troops into the area. This conflict, which was not declared as a war, continued until 1945 and proved the vital context for Japan's later wars.

THE SPANISH CIVIL WAR

Like international conflicts, civil wars could also show the role of the unexpected, and certainly not turn out as had been anticipated. In Spain, a group of right-wing senior army officers, who called themselves the Nationalists, sought to seize power in 1936 from the elected government, in a rebellion that began in Spanish Morocco on July 17 and on the Spanish mainland on July 18–19. They were opposed to the modernizing policies of the left-leaning Republican government. They were also concerned about the possibility of a Communist seizure of power via the *Frente Popular* (Popular Front) after the narrow left-wing electoral victory by the Popular Front in the hard-fought elections of February 16, 1936. The army's attitude to politics explains the rebellion by much of it. Claiming that the government had lost control (which in practice was due to right-wing as well as left-wing violence), they were really against the republic itself, and, with it, democracy and freedoms.

The Nationalists, however, achieved only partial success in 1936. They conspicuously did not take control of Madrid, let alone Barcelona. Moreover, the failure of any attempt by the rebels to negotiate a settlement or to defeat their opponents led to a bitter civil war that only ended on March 28, 1939, when the Nationalists seized Madrid.

The Spanish Civil War is commonly seen as a harbinger of World War II, and the ideological division between the two sides are emphasized. Indeed, the Nationalists, who very much emphasized religious themes, depicted the Republicans as the servants of the Antichrist. Caution, however, is required in seeing the Civil War as a harbinger of the World War, while the ideological dimension was scarcely new. It had played a role in the civil war aspect of the Peninsular War of 1808–13, and even more in the Carlist Wars. Conflict with Islam, a theme that was integral to Spanish history, identity, and ideology, was echoed and transformed by the Nationalist leader, Francisco Franco, into a conflict with modernity, notably "alien" Communism. Moreover, the practice of military violence was well established as a political tool in nineteenth-century Spain, as in Portugal and Latin America. In part, this was a baleful consequence of the destabilization wrought by the Napoleonic conquest in 1808. Against this background, the twentieth century offered a new iteration of old themes and methods, a situation also seen in the Balkans. Extreme violence was employed, as by the Nationalists in Seville in 1936, irrespective of the degree of violence required to achieve specific goals. Repression employing disproportionate force continued after particular places had been captured and, indeed, after the war. It was part of the war.

The Spanish Civil War also throws light on another aspect of the causes of war, namely the conflicts that did not break out. The combatants received extensive international support, but no formal war broke out as a consequence. This reflected a willingness and ability to fight by indirect means, a situation that raises questions about the definition of war. Italy and Germany sent substantial assistance to the aid of the Nationalists. Their forces that were involved were at war, but not formally waging it. So also, albeit to a lesser extent, with Soviet aid to the Republicans.

APPEASEMENT

The most controversial aspect of the causes of World War II relates to the argument that the British and French were partly responsible because of a failure to adopt a robust stance toward Germany, Japan, and Italy, the Axis powers, prewar. Voiced at the time, this argument was much employed by the Left during the Cold War that followed the war, in order to hold the West partly responsible for Axis policies and, indeed, World War II. This approach shifted the blame to Britain and France, a rather curious response to the goals and actions of the Axis. This approach is, to a degree, matched by the argument that much of the responsibility for the war lay with the global economic situation, namely with the sustained global economic Depression of the 1930s encouraging international competition and political support for extremists.

In practice, while systemic factors, such as the Depression, were highly significant in destabilizing the international system, Hitler's responsibility for the war was the key element. You do not blame a lake when a murderer drowns someone in this lake. Like other politicians, Hitler operated in response to a background that he did not create, as well as to international circumstances and developments. Nevertheless, Hitler also played a major role in shaping them and in encouraging a mistrust that made compromise appear a danger. He built on German anger with the territorial losses and reparations of the 1919 peace settlement (each of which Germany had imposed on France in 1871), but gave it a new inflection. The racial ideology and policy of the destruction of Jewry and the subjugation of the Slavs presented an agenda in which racial conflict was linked to an exaltation of violence. Ironically, as it sought to direct popular anger the Nazi Press Office was subsequently to attribute the outbreak of the war to the Jews, which was a classic instance of blaming the innocent, and one that mixed cynicism with paranoia. Subsequently, the Jews were again to be (inaccurately) blamed in Germany for Allied bombing, while (on a different basis but again blame-shifting) the Allies were to be criticized for not bombing the extermination camps.

The focus on Appeasement continues to play a significant role in current controversies. Aside from this specific argument about the origins of World War II, there was also the use of Appeasement outside this context, but as part of a call for action. A lack of deterrent firmness is presented as responsible either for war or for weakening the recourse for war. Thus, in 2003 and 2013, opponents of international intervention in Iraq and Syria respectively were described with reference to the Appeasers, as part of a long process of castigating caution.[7]

These lessons have frequently been applied in a far-fetched fashion, which demonstrates their malleability and resonance, as in the near-universal usage across the West of the Munich Agreement of 1938, which was the key episode in the Appeasement of Germany. Munich was also employed in 2013, by the Japanese when urging opposition to China's ambitions in the East China Sea, and repeatedly by Israel when pressing for opposition to Iranian nuclear plans. In 2014, it was employed anew when discussing the response to the Russian occupation of Crimea.[8] Historical analogy thus acted as a call to action, as it continues to do, and will continue to do. Rephrased, thinking by historical analogy is a key component of strategic culture. It is the choice of the relevant "lessons" that is crucial to the latter; there is scant inevitability about them.

The scholarly dimension is very different, for Appeasement emerges as in large part a matter of circumstances, notably, in the case of Britain, the interaction between far-flung imperial commitments and strategy. There is a corresponding emphasis on the extent to which British policy options were constrained by the need to protect threatened interests across the world, which prefigures the current American situation. The uncertainties affecting British policy related in part to this situation, but also to the extent to which it was by no means clear, prior to 1938, whether Nazi Germany or the Soviet Union was more of a threat. Furthermore, wherever the emphasis was placed, it was unclear how best to confront these threats.

The eventual outcome was far from predictable, a point that remains relevant today. In the case of the Soviet Union and Britain, there was hostility short of war in 1939 to 1941, then alliance against Germany until 1945, and then sustained opposition between the Soviet Union and Britain in the Cold War.

Some British and French commentators saw Germany as a potential ally against the Soviet Union. Moreover, Hitler initially hoped that Britain would join Germany in a war against Communism. However, in Britain, Hitler's determination to overturn the (much criticized and misrepresented) 1919 Versailles Peace Settlement,[9] and to make Germany a great power anew, was correctly regarded as a growing challenge to Britain's interests, which included the international order. In the winter of 1933–34, Nazi Germany was identified as Britain's ultimate potential enemy by the Defence Requirements Sub-Committee.

Germany was seen as a graver security threat than Japanese expansionism, even though the latter was already apparent in Manchuria. Britain's unwillingness to accept Japanese expansionism in China helped lead the Japanese navy in 1934 to begin preparing for war with Britain. This was a major step; the Japanese navy had developed on the pattern of the British navy, and with its assistance. Moreover, Britain and Japan had been allies from 1902, notably in World War I.

Focusing on Germany, Neville Chamberlain, Britain's prime minister from 1937 to 1940, made a major effort to maintain peace, and thus both domestic and international stability and the chance of economic recovery. However, Chamberlain was weakened by his inability to accept other points of view or to learn from experience, and by his self-righteous and continuing optimism about his own assumptions. Indeed, these flaws helped vitiate the conduct of British policy, ensuring that, however sensible in practice and/or as a short-term expedient, Appeasement was developed in a fashion that did not secure its purposes. Moreover, the implementation of Appeasement helps ensure that it is open to subsequent criticism, as did its presentation.

Chamberlain feared that war would lead to the collapse of the British Empire and would also wreck the domestic policies of the Conservative-dominated national government. He was indeed correct on both counts, although he was at error in seeing these outcomes as worse than the victory of Nazism. It was assumed that, if conflict broke out with Germany, then Japan might be encouraged to attack Britain's Asian empire, which was rightly seen as militarily and politically vulnerable. This vulnerability encouraged the British government to search for compromise with rising nationalism in India, not least with the Government of India Act of 1935, and with Ireland over naval bases in 1938. A sense of vulnerability also led to the attempt to create a viable policy of naval support, based on the new base at Singapore (opened in February 1938), for the British Empire in the Far East: Hong Kong, Malaya, Singapore, north Borneo, British interests in China, and links to the Dominions of Australia and New Zealand. However, the deployment of modern aircraft was focused on the challenge from Germany.

An American alliance did not seem a welcome solution; the Americans were regarded as posing a challenge to British imperial interests and, correctly, as unlikely to provide consistent support. This viewpoint can be difficult to recover from the perspective of subsequent wartime and postwar cooperation with America, both of which were crucial to Britain. Nevertheless, it is a viewpoint that helps explain the importance of this later cooperation. British responses to Japanese, Italian, and German expansionism were affected by the nature of Anglo-American relations, and, in turn, the legacy of these years helped underline later calls for a strong alliance. Isolationist America, which, under Franklin Delano Roosevelt, president from 1933 to 1945, had had cooler relations with Britain than those in 1929–31, and which passed Neutrality Acts from 1935, was regarded in the 1930s as self-interested. This, indeed, was a key element in American isolationism. Moreover, the two powers had failed in 1932 to cooperate against Japan during the crisis caused by its invasion of Manchuria.

Isolationist sentiment was strong in the United States, notably so from the reaction against President Woodrow Wilson and his role in the establishment of the League of Nations at the close of World War I. In 1937, this sentiment led Congress to consider the Ludlow Resolution, which would have required a national referendum before Congress could declare war, unless in response to a direct attack. That October, Roosevelt's "Quarantine" Speech, proposing that aggressor states be placed in quarantine, enjoyed only limited support in the United States in the face of strongly expressed isolationist views.

Such views were linked to a conviction that the United States should focus only on the defense of the New World—hemispheric defense—and, despite signs, such as the 1938 trade agreement with Britain, this approach helped make the United States a problematic potential ally. The situation was exacerbated by limited expenditure on the American military in the 1930s. This was not as bad as was later suggested. The oft-repeated comment that the army was smaller than that of Portugal is misleading; Portugal had extensive colonies in Africa to protect (and from which to raise troops), notably Angola and Mozambique, while, conversely, the United States spent much more on the navy and air force. Portugal, for example, had no equivalent to the thirty-five B17 Flying Fortresses deployed by 1941 at Clark Field, America's leading air base in the Philippines. Introduced from 1937, this was the first effective American all-metal, four-engine, monoplane bomber.

Nevertheless, the United States had a smaller military than it could afford as the world's leading economy and a major center of population. Crucially, the United States, in the 1930s, did not press forward with rearmament as its wartime opponents did, as well as the Soviet Union, France, and, indeed, Britain. In 1938, the American army could only put six divisions in the field, although it had one of the world's leading navies, an improving air force, and valuable developments in military planning. The United States could do far more than solely defend its own immediate geographical region. Moreover, during the subsequent world war, the United States benefited from not having earlier devoted massive expenditure to weapons that were out of date by 1940, notably with aircraft.

As far as Britain was concerned, Appeasement was designed to avoid both war and unwelcome alliances, and responded to the domestic mood. Britain in the 1930s certainly lacked a powerful alliance system comparable to that in World War I. Although hopes for the French defense of Western Europe in the event of German attack were high, France had been greatly weakened by World War I. It increased its military from 1935 in response to the German remilitarization of the Rhineland. However, France did not spend as much as a Germany solely focused on a military buildup, had a smaller population than Germany as a pool from which to recruit, as well as a smaller industrial base, had major colonial commitments, and also put strenuous and successful efforts into developing its navy, a navy in large part designed to act against Italy's new navy. [10]

Confidence in the ability of an Anglo-French alliance to prevent German expansionism in Eastern Europe was limited. Indeed, prior to the outbreak of a new war, Germany did well in the bitter competition for influence and markets in Eastern Europe that was a key aspect of the rivalry between the great powers. In seeking cooperation there, the Germans benefited from the "democratic deficit" across much of the region as well as from bitter opposition there to the Soviet Union. This opposition reflected the significance of ideological factors.

Moreover, unlike in 1914, neither Russia nor Japan was an ally of Britain. This absence was a key contrast, and even more so because both powers allied with Germany: the Soviet Union in 1939–41 and Japan throughout. However, despite significant economic assistance to Germany, the Soviet Union did not directly fight Britain, while Japan only did so from December 1941.

In the 1930s, the British government was unhappy about Britain's allies and potential allies. It was also unwilling to explore the path of confronting Hitler by making him uncertain about the prospects of collective action against Germany. Instead, as another instance of choice, the British government preferred to negotiate directly with the expansionist powers, especially Germany. This political response was matched by Chamberlain's focus on deterrence through a stronger navy and air force, each of which was to be based on Britain, rather than through an army that was to be sent to the Continent and therefore would have to cooperate with France. This buildup was an aspect of what was an unprecedented international arms race, unprecedented as it involved air power over both land and sea, as well as more conventional weaponry. [11]

The policy of negotiating with the Axis focused on Germany, because it was felt that Japan would be cautious if peace was maintained with Hitler. This was a reasonable view, at least insofar as Britain and its determination to preserve its colonies was concerned. It was certainly not so for China, which was the repeated victim of Japanese aggression. Italy, from 1922 under the bombastic and opportunistic Fascist dictator Benito Mussolini, was not treated as a serious threat, and, instead, was regarded by Britain as a possibly ally. Mussolini indeed long saw Hitler as a rival in Austria and the Balkans, although he shared both Hitler's contempt for the democracies and his opposition to Britain and France. These views became more important for him in the late 1930s with Hitler's growing power.

On the part of Britain, a sense that compromise with Germany was possible, combined with a lack of interest in the areas threatened by German expansionism, encouraged a conciliatory search for a settlement, as did the extent to which most were in denial about what Nazism was really like, both in domestic and international policy. In some respects, there was an attempt to reintegrate Germany into the international order that was comparable to the treatment of France after the Napoleonic Wars ended in 1815. Thus, Hitler was treated as another Napoleon III, the expansionist and bellicose ruler of France (first as president and then as emperor) from 1848 to 1870. Yet, such an approach was mistaken. The search for compromise with Hitler was not only unsuccessful, other than as a series of concessions, but also, arguably, discouraged potential allies against Germany.

Nevertheless, a problem with the postwar and current emphasis on the follies and failure of Appeasement is that, at the popular level, this emphasis can make it seem that Britain and France were in some way culpable, if not as culpable as the aggressors, and that blame should be divided. This assumption, which was particularly strong among those concerned to extenuate the Soviet Union, which actually allied with Germany, is deeply flawed. Ironically, this criticism often sits alongside attacks on Britain's wartime conduct, notably its area-bombing of Germany.

There is a parallel here with the treatment of the causes of World War I, as discussed in the previous chapter. Rather than treating all combatants as equally culpable, it is clear that, alongside systemic issues and tensions, particular responsibility attaches to the Austrian and German military elites. Britain, in 1914, acted against the unprovoked aggression and violation of international treaties represented by the German invasion of Belgium. The war was undertaken reluctantly by a people who quite genuinely believed in the value of peace and were sustained by the high moral value of self-sacrifice, and not by a crazed jingoistic mob.[12] Indeed, for Britain and the United States, World War I was a just and justifiable war. Moreover, that war did not end with a Carthaginian (harsh) peace, as, in contrast, the Germans were to claim. The 1919 Versailles peace settlement was far more generous than that after World War II involving as it did the occupation of all of Germany; but the latter peace did, and does, not receive the criticism directed at the former.

The attempt to contain the Axis powers short of war failed in the 1930s. Nevertheless, however mishandled, the realpolitik involved in Appeasement was not inherently dishonorable, although it is not clear that such latitude should be extended to those in the British government and establishment who advocated a negotiated settlement with Germany after the defeat of France in 1940. In the 1930s, a wish to seek a negotiated alternative to war was widespread across the political spectrum. Far from being a characteristic of reprehensible Conservatives, not to say fellow-traveling neo-Fascists, both the desire to avoid war and opposition to rearmament were also notably strong among liberal opinion and on the Left, and particularly so prior to the Czech Crisis of 1938. This point needs to be underlined because of the subsequent politicized placing of Appeasement, notably in Britain, as a means to criticize the Right.

As "Britain" as a term stands for the British Empire, it is also necessary to discuss the response of the Dominions: Australia, New Zealand, Canada, South Africa, and Eire (Ireland). Their attitude and effort had been crucial to the British effort in World War I,[13] not only in the fighting but also in the economic sphere. Yet, from the Chanak Crisis in Turkey in 1922 on, it had been clear that the Dominions were not only pursuing independent tracks, but also ones that could be at variance with those of Britain. In the Chanak Crisis, there was an unwillingness, apart from on the part of New Zealand and Newfoundland, to support Britain when Turkish nationalist forces under Kemal Ataturk approached the Dardanelles and Bosphorus, which were garrisoned by British forces in accordance with the postwar settlement imposed on Turkey in 1920. In the event, with most of the cabinet unwilling to risk war, Britain backed down and the 1920 settlement was revised in 1923. There was no comparable difficulty in political relations with the Dominions until the late 1930s. However, at the time of Appeasement, and notably the Czech Crisis of 1938, the unwillingness of the Dominions, especially Canada, to risk war was a key element.

Moreover, it is instructive to consider a figure from the political wilderness, not least because he reflected a strand of thought and had been a senior ministerial colleague of Winston Churchill, then also in the political wilderness. On his visit to Germany in the autumn of 1936, David Lloyd George, prime minister from 1916 to 1922, twice saw Hitler, whom he was particularly keen to assess. Lloyd George's visit led to much criticism; he was presented as being completely taken in by the dictator. The invitation originated with Hitler, who had praised Lloyd George's wartime leadership in his book *Mein Kampf* (*My Struggle*, 1925). The second meeting closed with Lloyd George urging Hitler, "the greatest German of the age," to visit Britain, and providing the assurance that "he would be welcomed by the British people." In practice, although Hitler wanted to enlist Britain against Communism, Lloyd George was not interested in taking good relations that far, and nor did he want them to compromise the Anglo-French alliance, which had played the central role in his wartime policy. Far from being a Fascist, Lloyd George can be located in the Liberal tradition, which was to seek good relations whatever the political complexion of the regime in question, as seen with the attitudes of Richard Cobden and William Gladstone toward Napoleon III of France. Of course, Liberals could also take a very different view, as, while in opposition, the hostile response to the Bulgarian Massacres in 1876 showed. Indeed, a week after his return from seeing Hitler in 1936, Lloyd George criticized Nazi political and religious repression: "a terrible thing to an old Liberal like myself."[14] Nevertheless, others were willing to overlook this repression in their enthusiasm for praising what was seen as a dynamic state that was apparently more promising than the capitalist democracies, although German dynamism in practice rested in part on debt.

Despite his concern about Hitler's internal policies, Lloyd George was to be convinced that an understanding with Hitler could have been reached. However, the Munich Agreement of 1938 was just such an understanding, and, despite his ambition to ally with the British empire, Hitler broke the agreement, as he would go on to break other understandings that restricted him. Indeed, this helped to give World War II a moral legitimacy as a "good" war with Hitler bearing the responsibility for the failure of Appeasement, which, indeed, conformed to the Augustinian ideal of seeking to avoid conflict. Hitler was interested, not in a revision of the Versailles settlement in pursuit of a German nationalist agenda, but in a fundamental recasting of Europe. This goal meant war, for only in defeat could an agreement that left the Continent at Hitler's disposal be seriously contemplated by Britain and France.

Moreover, Hitler was not a ruler with whom lasting compromise was possible, as Joseph Stalin, the Soviet dictator, discovered in 1941 when Hitler attacked his Soviet ally, which had provided Germany with crucial geopolitical and economic assistance in 1939–41. Hitler's quest for a very extensive, eventually continent-sized, empire reflected his belief in the inherent competitiveness of international relations, his system of aggressive racist geopolitics, and his conviction that widely flung territorial power was necessary for effective competition with the other empires or imperial-strength powers: Britain, the Soviet Union, and the United States. This was a conviction shared with Japan. Hitler wanted both to overthrow Versailles and, with increased urgency from late 1937, to redraw Eastern Europe. Ideally, this was to be on the lines, at least, of the victorious Treaty of Brest Litovsk negotiated with the Soviet Union in 1918 at a stage when Germany had done very well on the Eastern Front of World War I. In contrast, Hitler had less interest in Western Europe until 1940, and never showed the same commitment to territorial or ethnic changes there.

Violence was at the heart of Nazism. Violence was allegedly for the "good" of the "Race," and was seen as the ultimate purpose of life. This was a creed born of a violence that required the destruction of alleged enemies. All were linked: domestic dissidence, the Nazis claimed, had to be prevented in order to strengthen Germany for war. For example, the Third Reich's "state bishops" recast the Bible and produced an "official" version of the Sermon on the Mount that was a grotesque distortion. There were similar elements in the causes and nature of governmental support for Fascist or Fascist-style bellicosity and expansionism in Italy and Japan.

Returning to the judgement of Appeasement and, more specifically, the Munich Agreement, the total failure of the Western powers to protect Czechoslovakia against a Soviet-organized Communist takeover in 1948 provided, and should continue to provide, an opportunity to rethink more sympathetically the weakness of the Western position in 1938. Nevertheless, the opportunity for such a rethinking was not really taken, because the positioning of Nazi Germany and the Soviet Union (a wartime ally) in Western public opinion in 1948 was very different. In contrast, more positive academic reconsiderations of Appeasement gathered pace in the 1980s, in part as a result of a better understanding of the multiple problems facing Britain in the 1930s, but, in some cases, as a key aspect of a revisionist case that the war itself had dealt Britain and its empire a near-fatal blow, and thus that efforts to avoid it were commendable.

The latter argument came from certain conservative historians, such as John Charmley, but their thesis enjoyed only very limited political and public support. In large part, this lack of support was a product of the understandable widespread reluctance to abandon the standard account and to argue for the attempt to reach an understanding with Hitler. However, in part, the particular contours of the Right in the 1980s were the key issue, notably the successful call by Margaret Thatcher, prime minister from 1979 to 1990, for a robust opposition to the Communist bloc. Subsequently, it was argued that this opposition helped provoke the fall of the Soviet Union, and, by extension, that robust opposition to dictatorships was the appropriate course.

Thus, positive scholarly reevaluation of Stanley Baldwin and Chamberlain, the Conservative leaders and prime ministers who preceded Winston Churchill, and of the strand of Conservatism that he displaced, was not linked to Thatcherism in foreign policy. Indeed, her effort to increase ideological commitments, and to push the bounds of possibility, markedly contrasted with the 1930s' politics of prudence, and the resulting policies, notably Appeasement. The cult of Churchill, to whom Thatcher frequently referred as "Winston," was also important. For example, in the aftermath of the Falklands War, in a major speech given at Cheltenham on July 3, 1982, Thatcher quoted Churchill when arguing that Britain could be great again. Addressing the Czechoslovak Federal Assembly in 1990, she referred to the shame she felt as a result of the Munich Agreement.

This cult continued into the 2000s and 2010s. In 2001, Tony Blair, the Labour prime minister, presented President George W. Bush with an Epstein bust of Churchill that he kept in the Oval Office. In 2007, Gordon Brown, Blair's successor, at his first press conference with Bush, spoke of "a partnership founded and driven forward by our shared values, what Winston Churchill, who was the first British Prime Minister to visit Camp David, called 'the joint inheritance of liberty.'"

WAR IN EUROPE

Appeasement was a key issue in 1931–38, in the response to Japanese, German, and Italian expansionism. The situation changed on March 15, 1939, when Germany seized Bohemia and Moravia: the modern Czech Republic, bar the Sudetenland, which had been gained already by Germany as a result of the Munich Agreement, a loss that gravely weakened the Czechs' capacity to defend themselves. By this seizure, Hitler destroyed the Munich Agreement, the most public product of Appeasement, exposing its failure. This leaves to one side the argument that, in 1938–39, Appeasement "bought" time for rearmament, especially a marked strengthening of the British Royal Air Force (RAF). It was now clear that Hitler's ambitions were not restricted to bringing all Germans under one state, which was the case argued when his acquisition of Austria (1938) and the Sudetenland (1938) were discussed, and, indeed, also the German excuse when, on March 23, 1939, an ultimatum demanding the immediate return of the Memel (Klapeida) region was successfully issued to Lithuania.

In response to the destruction of the Munich Agreement, an attempt was made to create a collective-security alliance system that would contain Hitler through deterrence. This system, however, was weakened by the failure to include the Soviet Union. Again, as with Appeasement, and in some respects linked to it, there has been considerable debate over responsibility for this failure. Critics have argued that Anglo-French anti-Communism bore much of the responsibility. While that attitude was, indeed, significant, far more was involved. Again, this issue throws light on the more general problems of deterrence, particularly clashes of interest and views in conception and implementation.

First, Stalin was hostile to the West, indeed as hostile as he was to Germany, whereas Hitler was hostile to both the Soviet Union and the West, but more so to the former. Stalin's hostility to the West owed much to an opposition to democracy and bourgeois liberalism, as well as a legacy of hostility from the Russian Civil War on: Britain, France, the United States, and Japan had all then intervened against the Communists. The Soviet Union was far more hostile to liberal democracies than Romanov Russia, Britain's ally in World War I, had been. There was also an incompatibility in long-term goals between Britain and the Soviet Union, one that provided a key context for diplomatic relations, as had been the case with the failure of Anglo-Soviet naval armaments diplomacy in 1935–39.[15] As the world's leading imperial power, Britain opposed Soviet views and moves in a number of areas, notably China and Afghanistan.

A shared interest in revisionism and opposition to democracy, and a shared mentality of the "leader," provided a basis for agreement between Hitler and Stalin. Moreover, with their ideologies and practices of terror and violence, each appeared equals between whom an agreement could be made, indeed as a natural partner of the other. In the summer of 1938, Stalin had planned to approach Hitler for an alliance, in August 1939 they were to ally, notably against Poland, and, in early 1940, there was to be pressure in Britain and France to send military assistance to Finland, a neutral power that had been attacked by the Soviet Union. There are major archival problems in studying Soviet policy, with limitations on access to sources and concerns about their reliability,[16] but benign accounts of Stalin's intentions are deeply suspect.

Second, serious differences between Poland and the Soviet Union played a key role in 1939. Negotiations for a triple alliance of Britain, France, and the Soviet Union, an alliance that might have deterred Hitler from acting in 1939, collapsed largely because Britain and France could not satisfy the Soviets on the issue of Polish and Romanian consent to the passage of Soviet forces in the event of war with Germany. Romania was suspicious about Soviet intentions about Bessarabia, a region seized by Russia from the Turkish empire as a result of war in 1806–12, that Romania had gained after World War I. Events were to vindicate this suspicion. Romania was also fearful of the German reaction if it agreed to the British approach.

Earlier, the idea of a four-power declaration by Britain, France, Poland, and the Soviet Union had fallen foul of Polish opposition. In light of the Soviet invasion of Poland in 1920 and treatment of Poles in the Soviet Union during the 1930s Purges, as well as what was to happen in 1939–41 and 1944–45, with the Soviets brutally occupying much and all of Poland respectively, Polish concerns are understandable. However, the Poles were also naïve in imagining that Germany could be restrained without active Soviet assistance. Stalin was totally untrustworthy, but the Poles had no viable alternative bar becoming a German client-ally. There were parallels between the Polish response and the divisions between the Baltic States in their response to the serious threats from Germany and the Soviet Union, divisions that weakened any attempt to build up a league. Whereas the Soviet Union was the prime enemy for Estonia, Germany was for Lithuania, while Latvia saw both as enemies. Similarly, Estonia and Latvia viewed Poland as an ally, while Lithuania had a bitter territorial dispute with Poland over the city of Vilnius and the surrounding area. This region symbolized the complex and rival strands of identity frequently seen in Eastern Europe, strands that were substance as well as shadow. Polish forces had taken Vilnius from Lithuania in 1919.[17]

The German-Soviet Pact of August 23, 1939, named after their manipulative foreign ministers Ribbentrop and Molotov, who negotiated the agreement, was crucial in encouraging Hitler to invade Poland, given the unexpected (and implausible in the light of their policy in 1936–38) determination of Britain and France to fulfill their guarantee of Polish independence, and the failure of his major ally, Italy, to act in Hitler's support. Hitler persisted against Poland despite the guarantee. He believed that Britain and France would not fight, especially as a result of his pact with Stalin. This freed Germany from a two-front war,[18] the disastrous situation that had faced Germany for most of World War I. However, the pact greatly alienated Germany's ally Japan, notably because Japanese and Soviet forces were involved in a major clash in August 1939. Thereafter, Japanese policy makers did not feel that they could trust Hitler, a situation that looked toward very different policies toward the Soviet Union in 1941.

In 1939, the British Chiefs of Staff advised that it would not be possible to offer Poland any direct assistance. This was to be borne out by the course of the war in both 1939 and 1944 and was also true of Britain's military posture toward Czechoslovakia in 1938 and 1948. Britain itself was left reliant in 1939 for military action in Europe on the French, who were to be revealed, when Germany attacked in 1940, as a flawed ally, both militarily and politically. Already, in 1939, France, despite its large army, was unable and unwilling to attack Germany's western frontier. Such an attack could only have been an indirect help to Poland (important as that would have been), whereas, with a long common frontier, the Soviet Union could readily have sent forces to Poland's assistance.

In the event, cooperation between Germany and the Soviet Union was initially directed against Poland and was also linked to an anti-Western turn in German policy. Indeed, Germany's war in the West was an ideological struggle like that for *Lebensraum* in the East.[19] Although crucial to the diplomacy and warfare of 1939–41, the German-Soviet Pact became an aspect of history written out of the Soviet account and, indeed, of that of Communist parties and leaders who praised it at the time, such as Mao Zedong. The pact had led to great suspicion of Communists at the time. The French government clamped down on the Communists. Insofar as it was discussed subsequently, the pact was misleadingly presented by apologists as an opportunity to gain time and space to resist German attack, not least by moving Soviet defenses forward. The Left has always treated the pact as producing Soviet neutrality. In fact, it was an alliance and perhaps without it there would not have been the millions of Soviet dead in the war. The pact, which divided Eastern Europe into German and Soviet spheres of influence and expansion, was subsequently frequently referred to by external and internal opponents of the Soviet system. For example, the fortieth anniversary of the pact in 1989 was marked by opponents in the Baltic Republics, as an aspect of the discrediting of the Soviet Union.

The outbreak of war on September 1, 1939, was not the only conflict in Europe that has to be explained as part of World War II. Far from seeing a process of inevitable expansion of the conflict, it is important to underline the extent to which the expansion of the war was hesitant. Moreover, this hesi-

tancy provided opportunities for the Axis powers. Crucially, Hitler's successive attacks from 1939 to the spring of 1941 found the Soviet Union neutral,
complicit and in practice allied (and an active participant against Poland
from September 17), and, very differently, the United States unwilling to
come to the aid of fellow neutrals. Thus, as in the Wars of German Unification (1864–71), Hitler was able to fight single-front conflicts. Film provided
historical endorsement for Hitler's diplomatic policy, with *Bismarck* (1940)
showing the firm and iconic chancellor, the maker of German unification and
a model of leadership on the Right, favoring a Russian alliance to protect
Prussia's rear.

Moreover, when Hitler invaded in 1939, Romania did not aid Poland,
while Norway and Sweden provided only limited assistance to Finland when
it was attacked by the Soviet Union in the Winter War of 1939–40. In
Norway, the government proclaimed neutrality and forbade all officers and
conscripts from taking part on the Finnish side, although it did send weapons
to Finland. There were also several aid campaigns in Norway where food and
clothes were collected and sent to Finland. Some men volunteered to fight on
the Finnish side, including some who were on the Far Right, seeing this as a
double opportunity to fight for a neighbor and against Communism. In Sweden, the support for Finland in the Winter War was stronger. The Swedish
government declared neutrality but allowed its troops to volunteer to fight for
the Finns, while there was also considerable support beyond the military.
However, if there was significant sympathy for the Finns, realpolitik prevailed at the governmental level. Crucially, Norway and Sweden refused to
allow transit for any Anglo-French relief force, a major factor discouraging
an intervention that was planned but that would probably, anyway, have been
unsuccessful. As a consequence, war did not break out.

Similarly, Japan was able to attack China from 1931 without the intervention of the other powers (including the neighboring Soviet Union), although
they were not otherwise engaged in warfare. Japan was also able to fight
limited border wars with the Soviet Union (1938 and, more seriously, 1939)
that did not escalate, and to attack Britain and the United States in 1941
without fear of Soviet entry into the war. When Roosevelt, Churchill, and
Jiang Jieshi met in Cairo in November 1943, Stalin was not present; he did
not wish to attend a conference of the powers at war with Japan. As a result,
he was not a party to the Cairo Declaration of December 1, by which Roosevelt and Churchill agreed to support the return to China of Taiwan (lost to
Japan in 1895) and Manchuria (lost in 1931–32). In contrast, Stalin was
willing to meet Roosevelt and Churchill at Teheran in December 1943; Jiang
was not present.

Yet, despite the reluctance of most of the powers to fight, Hitler's inability to consider limits on his ambition, or to understand the flaws in German
economic and military capability, acted as an inexorable factor encouraging

the expansion of the war. A correct sense that Hitler could not be trusted also meant that the rapid defeat of Poland did not lead to negotiations for peace. In the *Reichstag* (Parliament), on October 6, 1939, Hitler called for peace with Britain and France, but no real attempt was made to compromise with them, let alone with conquered Poland. Moreover, Britain and France were determined to fight on to prevent German hegemony in Europe and planned for success through a long war involving naval and economic blockade and the active pursuit of alliances. There was also the hope that Hitler would be overthrown. The strength of the British war machine and the global range of the British economy were significant to this planning.[20] It provided a measure of confidence in victory that encouraged persistence in a struggle for which, in practice, there was no clear way out.

NEW COMBATANTS IN 1940

While the German invasion of the Netherlands and Belgium on May 10, 1940, was an aspect of the plan for the conquest of France launched the same day, there was no such need for the German invasion of Norway on April 9, 1940, an invasion that, in turn, required that of Denmark the same day. The Japanese were to adopt a similar logic of attack on successive dominoes in their drive south in late 1941. A determination to break the British naval blockade, and to give the German navy operational flexibility and strategic possibilities, played a major role in German policy against Norway.[21]

In the event, the conquest of Norway brought relatively little benefit to German war-making, in part because serious weaknesses in the German surface navy made Norwegian bases less valuable than might otherwise have been the case, while the German conquest of France swiftly provided submarine bases closer to Atlantic trade routes. Brittany indeed was so positioned as if designed to threaten British maritime links. Furthermore, concern about the possibility of British attack ensured that the Germans tied up a large force in garrisoning Norway, a force that, in the event, had no strategic value, as there was no comparable risk of it being used to invade Britain. This force was not needed to coerce neighboring Sweden, which was very willing to trade with Germany. Similarly, the German conquest of Yugoslavia in 1941 led to an even larger garrison obligation because of the serious problems posed by guerrilla opposition there. This opposition encouraged British intervention in favor of the guerrillas.

This point underlines the problematical nature of cost-benefit analyses of the value of going to war, a process anyway in part dependent on the nature of the regime in question and of its ideology. There is scant investment within a military or militarized regime in arguing for the folly or weakness of a reliance on force; although failure can lead to its fall, as with Argentina and the Falklands War in 1982. Planning the invasion of Yugoslavia in 1941, the Germans did not consider the possibility of large-scale post-conquest resistance and the need subsequently to deploy large numbers of troops there accordingly.

The neutrality of the Netherlands and, more particularly, Belgium and their failure to create a joint defense,[22] gravely weakened the Anglo-French response to German attack in 1940. The failure to allow the prior deployment of British and French forces in these countries meant that these forces had to advance and to expose themselves to the German attack without the benefit of holding defensive positions. The rapid and total success of this attack was a striking contrast to the situation in World War I; and the 1940 campaign culminating in the fall of France in June was important to the politics of the war, both those at the time and those subsequently. The outcome of the campaign also permitted a vindication of Germany's decision for war the previous autumn. The failure of Allied deterrence appeared to rest on a sensible German assessment of the situation, an approach that underplayed the major risks taken.

WAR WITH VICHY

With France's surrender in June 1940, there was no longer the prospect of a main-force conflict with the Germans on the Western Front, as in World War I. Instead, there was now the need to limit the consequences of France's surrender and of Italy's entry into the war. A key characteristic of Churchill's strategy was a determination to attack. He responded to the threat that the French navy, now in the hands of the Vichy regime that was aligned with Germany, would be taken over by the Germans, or would cooperate, by ordering action against it. This turn to war remains a controversial step.

Moreover, driving the Vichy French from their possessions was regarded as a crucial way to win the global struggle for power. On September 18, 1940, Churchill wrote to Sir Samuel Hoare, the ambassador to Spain, explaining his support for the Free French attempt to capture Dakar, the capital of Senegal, the leading French colony in West Africa. Churchill presented a classic account of the indirect approach, observing that, if Charles de Gaulle were to establish himself in Dakar and become "master of Western and Central Africa, Morocco is next on the list."[23] The struggle was to broaden out to encompass Vichy possessions elsewhere, notably Syria (1941), Lebanon (1941), and Madagascar (1942), each of which was conquered by British-led Allied forces.

ITALY ENTERS THE WAR

In June 1940, the Italian dictator, Benito Mussolini, an ally of Hitler in the Pact of Steel, joined in the war. He did so once the French had been clearly defeated by the Germans, yet while the conflict with France and Britain was continuing, because he feared that he would otherwise lose the opportunity to gain glory and territories from France and Britain. Longer-term ambitions also played a major role. German victory in 1940 brought to a head Mussolini's ideological affinity with Hitler, his contempt for the democracies, and the inherent violence and aggressive expansionism of his regime and overcame his awareness of the poor financial and military situation of Italy and his realization of little popular support for war. At the same time, Mussolini's awareness could be overcome by his delusions.

Mussolini had already taken a very brutal role both in suppressing resistance in Italy's colony of Libya and, in 1935–36, in the conquest of Ethiopia, which was then called Abyssinia. This conquest flew in the face of the League of Nations' commitment to a peaceful settlement of disputes, but Britain, despite preparations, and France did not intervene against Italy. On April 7, 1939, Italy pressed on to invade Albania, an independent state where Italian influence was already strong. Albania rapidly fell, and Victor Emmanuel III of Italy was declared king of Albania.

War with Britain and France was conflict in a very different league for Italy, matching the process of escalation also seen with Germany and Japan. Feeling that his vision of Italian greatness required domination of the Mediterranean and, therefore, British defeat, Mussolini, who had built up a large and expensive modern fleet to that end, sought gains from France and the British Empire. This policy, which entailed overcoming the pronounced pessimism of his service chiefs, destroyed the conviction in the British Foreign Office that Italy would remain neutral and the linked illusion that Mussolini could be a moderating "back-channel" to Hitler. This had been hoped by those on the Right in Britain keen to see a negotiated end to the war.

Italian expansionism in the late 1930s and 1940 represented a continuation of Mussolini's breach with both Britain and France and, instead, alignment with Germany as a result of the Ethiopian Crisis of 1935–36. In November 1938, the Italian Chamber of Deputies had echoed to calls for acquiring "Tunisia, Corsica, Nice, Savoy,"[24] all from France. Mussolini judged Corsica, purchased by France from Genoa in 1768, as traditionally Italian,[25] but did not have the same view of Savoy.[26] In his forceful report on February 4, 1939, to the Fascist Grand Council, he showed a clear opposition to France. Even though relations were close with Hitler, notably with the Pact of Steel signed on May 22, 1939, this depended on a verbal understanding that neither Germany nor Italy would provoke war before 1943.

Mussolini's son-in-law and foreign minister, Count Galeazzo Ciano, opposed this alignment. He claimed that Mussolini was furious with Hitler for not consulting him about the invasion of Poland. Mussolini appreciated that Hitler, rather than being Mussolini's equal, was becoming more prominent. The invasion of the Low Countries and France made the envious Mussolini conclude that he could not remain neutral. He was convinced that war would be short and that Germany was powerful enough to win.

Assisted by the alpine terrain, by prepared defenses, and by a lack of Italian preparedness for launching an offensive, the French mounted an effective resistance to Italian attacks in June 1940, and, in the subsequent peace, signed on June 24, Italy's territorial gains were very modest. These terms reflected the success of the French resistance to Italian attack, but also Hitler's concern to bolster Vichy France.

Whatever the arrangements that would have been offered by Germany had Britain accepted peace (i.e., surrendered) in 1940, Britain fought on. Aside from the subsequent air war over Britain (the Battle of Britain and then the Blitz) and the U-boat (submarine) assault on its trade, most of the initial fighting on the part of the Axis involved Italy because the projected German invasion of Britain, Operation Sealion, was not launched. Although vulnerable British Somaliland fell rapidly, Italian campaigning in East and North Africa was a humiliating failure. Mussolini's planned advance on the Suez Canal, the vital axis of British imperial power, failed totally. In turn, in 1940–41, Somalia, Eritrea, Ethiopia, and much of Libya were captured by British troops (many from the empire). This failure led to a major shift in the politics of the Axis, with Germany ever more dominant.

The same shift in Italian policy, from ambition to action, occurred in the Balkans. Mussolini's reckless ambition for Balkan gains led Italy to attack Greece in October 1940. The two powers had had poor relations from the 1910s, over rival territorial interests in the Balkans and the Aegean. Thanks to the Italian colony of Albania, they also had a common frontier. In turn, Italian failure in this conflict, which resulted in a successful Greek invasion of southern Albania, ensured that Hitler decided to intervene in Greece. This operation, however, when launched on April 6, 1941, was an attack on both Greece and Yugoslavia, because an unexpected nationalist coup by the military in Yugoslavia on March 27, encouraged by Britain, challenged German influence there and Hitler was not prepared to accept such a challenge.

SOVIET EXPANSION

A focus on German, Italian, and Japanese expansionism in 1939–41 can lead to an underplaying of that of the Soviet Union, which, instead, sought to keep pace with that of Germany, not least to give effect to the division of Eastern

Europe agreed in the Nazi-Soviet Pact in August 1939. There is a tendency to underplay this expansion, but for Poland, Finland, the Baltic Republics (Estonia, Latvia, Lithuania), and Romania, it was a brutal process. The Soviets only annexed part of Finland and part of Romania, but in eastern Poland (overrun in 1939) and the Baltic Republics (1940), their complete control was accompanied by the large-scale deportation to the *gulags*, the often-deadly Soviet concentration camps, of large numbers of those judged potentially hostile.

To the Soviet Union, this expansion was in part a matter of re-creating the pre-1914 Russian empire, which had included all these areas; rather as the Nationalist (Guomindang) government of Chiang Kai-shek sought to do for China in Xinkiang, where (after autonomous government by a warlord, Sheng Shi-ts'ai) Nationalist control was reasserted in 1943. The Nationalists also sought to reassert control in Tibet, the latter a challenge to British support for Tibetan autonomy and thus for a buffer to British India.[27] For those affected by Soviet expansion, the situation was very different: sovereign states whose independence was internationally recognized were forced to surrender independence (the Baltic Republics and, in concert with Germany, Poland), or territory (Finland and Romania).

GERMANY ATTACKS THE SOVIET UNION

The causes of the war between Germany and the Soviet Union in 1941, launched by Germany on June 22, the largest-scale land conflict of World War II, can be firmly found on the German side. There have been repeated claims that the Germans preempted plans for a Soviet attack, a claim initially made by Josef Goebbels, Hitler's minister of propaganda, in June 1941. Whatever the nature of long-term Soviet intentions and military expansion, and Stalin might well have attacked after 1943 if Germany had been weakened by its conflict with Britain (and maybe the United States), this claim was not an accurate account of the situation in 1941.[28] Instead, Hitler's overconfidence and contempt for other political systems reinforced his belief that Germany had to conquer the Soviet Union in order to fulfill what he alleged was her destiny and to obtain *lebensraum* (living space).[29] The earlier reversal, from 1933 to 1940, of the Treaty of Versailles, was secondary to the reestablishment of Germany's victory in the East in 1918, a victory rapidly overthrown as a consequence of Germany's defeat later that year on the Western Front.

Such a war would also permit Hitler to pursue the extermination of the European Jews that he had promised the *Reichstag* on January 30, 1939. To Hitler and his supporters, "International Jewry" was an active worldwide force, responsible for Germany's plight, and therefore had to be destroyed in order to advance the German cause.[30] Hitler's meta-historical goal of racial superiority, especially over Slavs, and of the slaughter of all Jews, was not an outcome possible without a total victory on the pattern of the pyramids of skulls left by certain medieval Asian conquerors, such as Timur (Tamberlane), when punishing opposition. At the same time, success encouraged Hitler to make his ambitions more central, not least to move from an emphasis on revising Versailles, so as to ensure a Greater Germany, to creating, instead, a new European order, one in which there would be no Jews.

Hitler was convinced that a clash with Communism was inevitable, as well as necessary. Soviet pressure, notably during the Molotov-Ribbentrop discussions in Berlin in November 1940 over the future of the Balkans, encouraged this view. Although Britain in 1941 was still undefeated, and Germany deployed considerable forces in France and Norway, Hitler correctly felt that Britain was no longer able to make any effective opposition to German domination of mainland Europe, and therefore that her continued resistance should not deter Germany from attacking the Soviet Union. Indeed, Soviet defeat was seen as likely to weaken Britain, by ending hopes from a war between Germany and the Soviet Union, and thus to encourage Britain to negotiate.

Furthermore, indicating the different timelines of anticipation that play a role in the causes of war, Hitler was convinced that the United States would enter the war on Britain's side. Indeed, on August 25, 1941, he told Mussolini that Roosevelt was controlled by a Jewish group, a totally inaccurate view similar to his flawed assessment of Churchill. The influence of Jews in the Soviet Union was also greatly exaggerated. Hitler believed that an attack on the Soviet Union was necessary in order to win rapid victory before such an American intervention.

Hitler's adventurism and conceit were a reflection of his warped personality and also the product of a political-ideological system in which conflict and hatred appeared natural. Moreover, Hitler was confident that the Soviet system would collapse rapidly and was happy to accept misleading intelligence assessments of the size and mobilization potential of the Red Army. German optimism was enhanced by the successful conquest of Yugoslavia and Greece in April 1941.

Equally distrustful of Britain and suspicious of plans for an Anglo-German peace, Stalin totally ignored warnings from German opponents of Hitler via Soviet intelligence. Warnings that were ignored also came from Roosevelt, and from Churchill (derived from Enigma decrypts), about German invasion plans.[31] As also with Hitler in 1941, this was a key instance of a war leader's willfulness, which indeed is a theme of the book, willfulness linking the context and goals of policy to its implementation.

German assessments drew on the flaws revealed in the Soviet invasion of eastern Poland in 1939 and, even more, in the initial stages of the Soviet attack on Finland in 1939–40, the "Winter War," as well as on assumptions based on the effect of Stalin's extensive purges of the military leadership from 1937 to 1941, purges themselves fed by the German provision of forged information.[32] These assessments failed adequately to note the success of the Red Army in the last (successful) stage of the "Winter War" as well as postwar improvements, while, anyway, there was a major contrast between weaknesses on the offensive and capabilities in defense.[33] Furthermore, Russian military defeat and total political collapse in World War I in 1917 encouraged the Germans to feel that victory could again be had. Letters indicate that many German troops were confident that the war would be over speedily. Nevertheless, despite their misjudgment of the Red Army, appreciable casualties were anticipated by the Germans, who deployed as many troops as they could. The Germans did not understand the Soviet ability to compensate for casualties and lost equipment, and notably the potential of the Soviet economy, especially in tank production. They also planned to overrun production facilities fast, thus making them unimportant.

There is also the question of the German way of war, and whether a national culture of war-making, with its emphasis on short wars characterized by offensive operations designed to lead to total victory, encouraged the resort to war and, indeed, helped cause German failure.[34] Academic interest in national cultures of war-making, and in national strategic cultures, opens another field for discussion of the politics of the war because the political cultures of individual states were bound up with their distinctive goals and types of war-making. Most German generals were confident that they had developed a military system able to defeat the Soviet Union. They did not understand Soviet geography, including the impact of the weather on the roads.

GERMANY'S ALLIES

Allies of Germany were expected to support the attack on the Soviet Union, and indeed did so, although for their own reasons. The Finns were determined to regain the territories they had lost to the Soviet Union in 1940, and saw what they termed "the Continuation War," of 1941–44, as a second stage of, and response to, the Soviet attack in the "Winter War" of 1939–40. The same goal of recovering territory was sought by Romania, which aimed to regain Bessarabia and Northern Bukovina from the Soviet Union. Thus, Romania wished to reverse the consequences of the Nazi-Soviet Pact. In order to help overcome Communism, and under pressure from Germany, Hungary declared war on the Soviet Union. Hungary's government had an explicitly Christian nationalist ideology[35] and was, by the early 1940s, moving the country toward becoming a racial state.[36] Moreover, in 1938–41, Hungary had benefited greatly territorially from alliance with Hitler at the expense of Czechoslovakia, Romania, and Yugoslavia, recovering losses under the Treaty of Trianon of 1920.

Yet, as is inevitably the case, these explanations are far too brief. To take Romania, there was a complex interplay between King Carol II, the Fascist Iron Guard, liberal politicians, and General Ion Antonescu, the minister of war, an authoritarian nationalist seen as above politics—with the Germans also playing a major role. Carol called in Antonescu as prime minister in 1940 as a way to contain the Iron Guard, and Antonescu, a nationalist who did not want to be client of Germany, nevertheless turned to Hitler; he was convinced that Germany was going to win the war, which indeed seemed a reasonable conclusion in September 1940. In turn, although the SS backed the Iron Guard, the German government, influenced by its army and the Foreign Ministry, supported Antonescu because they saw him as a reliable ally and a source of stability, neither of which was true of the Iron Guard. In light of the forthcoming war with the Soviet Union, reliability was more desirable than the ideological affinity offered by the Iron Guard. At the same time, Antonescu and his government were to be murderously anti-Semitic.

Underlining the interaction of domestic politics with foreign policy, Antonescu bullied the unreliable Carol into abdicating in favor of his son, Michael, on September 6, 1940. The following January, Antonescu saw Hitler, telling him that he was ready to support Germany in defense of Eastern Europe against a possible Soviet attack, and, eight days later, Antonescu was able to crush the Iron Guard without German intervention on behalf of the latter. In June 1941, Romania joined in the attack on the Soviet Union, declaring a "holy war" to free Bessarabia. Antonescu saw Germany and the Soviet Union as the only alternatives.[37] This was an analysis that France's defeat and British weakness encouraged many to share.

WAR IN THE PACIFIC

The outbreak of the Pacific War in December 1941 is harder to study because, partly due to the nature of the surviving sources, Japanese policy making has proved more opaque than that of Germany: it is less easy to see the precise role of individuals in discussions. Nevertheless, the general situation is clear. Japan was already at war with China. The wider conflict begun in 1941 amounted to more enemies to beat and was not the much bigger step to go from a peaceful to a wartime society.

Alongside an expansionism based on the self-confident assumptions of the Japanese ruling elite, resources issues, particularly access to oil, were important precipitants to the decision for war.[38] These issues played a key role in clashing geopolitical priorities focused on Southeast Asia. The collapse of France and the Netherlands to German attack in 1940, and the weakening position of Britain, already vulnerable in the Far East (of which the Japanese were well informed), created an apparent power vacuum in East and Southeast Asia. This vacuum encouraged Japanese ambitions southward into French Indochina (Cambodia, Laos, and Vietnam), British Southeast Asia (Malaya, Singapore, north Borneo, Burma), and the Dutch East Indies, while leading the Americans to feel that only they were in a position to resist Japan. The last was accurate.

Jiang Jieshi, the Chinese leader, had argued in his diary on September 2, 1939, that it was crucial to link the Sino-Japanese War with the Allied cause in Europe.[39] Although this did not happen in the way he predicted, Jiang was correct in feeling that, to move beyond checking Japan, it was necessary for China to benefit from developments in the international system.

Furthermore, the unresolved character of the war between Japan and China, which entered a state of attritional stalemate in 1939,[40] not only embittered Japanese relations with America, which provided some support for China, but also exacerbated resource issues in Japanese military planning, as well as placing a major burden on Japanese finances. Thirty-four divisions were bogged down by 1939; by that September, Japan had lost half a million troops killed or badly wounded in China. Japan had had to settle on a policy of consolidating the territory it controlled in China and launching punitive expeditions into the remainder of China. Moreover, Japanese economic weakness was such that her trade deficit was condemning her to the prospect of national bankruptcy in the spring of 1942. This was a repetition of the serious fiscal and economic strain that had confronted Japan during the militarily successful and far shorter Russo-Japanese War of 1904–5, but it had proved possible to end that conflict more speedily.[41] There was no reason to believe that the war with China could be ended so readily. The failure of Japanese policy makers was clear but unacknowledged.

Germany's victory over France in 1940 encouraged Japan to revive relations with Germany that had been dimmed when Hitler, without warning, concluded his nonaggression pact with Stalin the previous year: Japan was opposed to the Soviet Union for ideological and geopolitical reasons and had sent a large force into the Soviet Far East during the Russian Civil War as well as briefly fighting the Soviet Union in Mongolia and Manchuria in 1938–39. On September 27, 1940, in response to pressure from the military, Japan joined Germany and Italy in the Tripartite Pact. The Japanese government and military, although divided, were determined to expand at the expense of others, particularly, from 1940, into Southeast Asia, which was, to Japan, the "Southern Resources Area," a region rich in raw materials, notably oil, tin, and rubber, all of which Japan lacked. From June 1940, the Japanese navy had begun full mobilization, although the German failure to invade

Britain in the autumn of 1940, as planned, discouraged ideas of a Japanese attack on Britain's colonies that year.

Continued Japanese aggression against China and, more particularly, expansion into southern French Indochina in July 1941 helped to trigger American commercial sanctions, specifically an embargo on oil exports to Japan. This was tantamount to an ultimatum because, without oil, the operations of the Japanese armed forces, notably the navy, would grind to a halt. Unlike the United States, a major exporter as well as producer, Japan did not produce oil.

Prefiguring recent debates over the role of oil in international relations, the ultimatum helped provoke the Japanese to act against the United States to protect their position and potential, because they were unwilling to accept limitations to their expansion in the Far East. The latter was more an "idealist" than a "realist" position. There was no inevitable need for Japan to confront the United States in this fashion.

In 1941, the Japanese increasingly focused on the raw materials to be gained from Southeast Asia and the East Indies. They also planned to seize British-ruled Burma (Myanmar) in order to block Western supplies to China, a goal that had already led them to occupy the northern part of French Indochina in September 1940. It was misleadingly believed that blocking these supply routes would lead to the serious weakening, if not collapse, of Chinese resistance. The Dutch East Indies (modern Indonesia) posed a problem for Japan as, despite the German conquest of the Netherlands in May 1940, the Dutch colonial officials rejected Japanese efforts to acquire oil, and instead sought to align policy with Britain and the United States.

Supplies for Japan and China had become far more of an issue with the development of American policy. The Americans considered themselves entitled to react forcefully to events on the other side of the Pacific, and also felt threatened by the fall of France in 1940 and by the possibility that Britain would follow, thus completely exposing the Atlantic to German action and the United States to attack by a worldwide coalition.[42] These anxieties encouraged Congress to support a rearmament made necessary by the weak state of the American military, and also led Roosevelt, in part to demonstrate bipartisanship, to appoint Henry Stimson as secretary of war on June 19, 1940. Stimson, a prominent Republican who had served as secretary of war under President Taft (1911–13) and secretary of state under President Hoover (1929–33), argued that American security required the maintenance of British power. The Neutrality Act was repealed, military equipment was sold to the British, and steps were also taken to confront Japan.

At the same time, there were strong divisions in public opinion and public politics. The isolationist America First Committee was formed in September 1940 to oppose intervention against Germany. The Committee to Defend America by Aiding the Allies and, later, the Fight for Freedom Committee took a very different stance.[43] The Two-Ocean Naval Expansion Act was passed on July 19, 1940, when Britain appeared defeated by Germany and certainly could no longer rely on the French navy. This act, increasing the authorized tonnage of American warships by 70 percent and providing for a cost of $4 billion, served notice on the Japanese that the Americans were going to be in a position to dominate the Pacific in the near future. This fleet would enable the Americans to wage naval war with modern ships against both Germany and Japan, a necessity that seemed increasingly apparent.

This buildup had an impact on Japan comparable to pre-1914 German fears of Russian military developments: the vista appeared threatening, but, in the short term, a window of opportunity seemed to be present. In turn, although it did not seek war with Japan, the American government was resolved to prevent Japanese expansion, yet unable to make an accurate assessment of Japanese military capability. The latter issue was linked to an American exaggeration of the effectiveness of their air power.[44] There was no experience of long-range Japanese naval and amphibious capabilities, especially the ability to mount multiple landings at the same time over a large distance, and a degree of racial discrimination in the assessment of the Japanese.

French Indochina, especially Vietnam, came to play a crucial part in the crisis. America registered responses to what she saw as aggressive Japanese steps there. After the fall of France, Japan ordered the closure of the border between China and Indochina to prevent the movement of supplies to China. Indochina was also of strategic importance as an axis of Japanese advance to the "Southern Resources Area." From air bases in Vietnam, it was possible to threaten Thailand and the British colony of Malaya, a point that anticipated later American concerns about Vietnam in Communist hands. The declaration, on July 24, 1941, of a joint Franco-Japanese Protectorate over all French Indochina led to an American trade embargo and the freezing of Japanese assets in America. Under the Hull Note of November 26, a memorial presented by Cordell Hull, the secretary of state, the Americans demanded that the Japanese withdraw from China and Indochina. Long-standing Japanese fears of encirclement now focused on anxiety about the so-called ABCD group: America, Britain, China, and the Dutch.

Within Japan, there were attempts to explore the idea of better relations with Britain and the United States, but these required an ability to restrain the military that did not correspond to the dynamics of Japanese politics. These explorations, already seen in 1939–40, were pursued in late 1941 in order to ascertain if the United States could be persuaded to lessen its support for China. Britain and China were anxious about this point, but, in November 1941, Roosevelt agreed that the United States could not sacrifice China in order to maintain relations with Japan. Thus, a Japanese recognition of Chinese independence became a key American objective, continuing a pattern of international restraint on Japanese expansion there seen episodically from the mid-1890s. To Japan, this was American intervention in its sphere.

After staging war games in August 1941, the Japanese decided to launch a war if diplomacy failed to lead to a lifting of the trade embargo. On October 17, 1941, a hard-line ministry under General Hideki Tōjō gained power. At the same time, as a reminder that it is necessary to appreciate the divisions and strains within states and their governments, Tōjō rose to power against the background of long-standing tension within the army, notably between the Control and Imperial Way factions, as well as within the government. There were disagreements over policy, especially whether there should be compromise with the Chinese Nationalists, an option opposed by Tōjō; and also over the Neutrality Pact signed with the Soviet Union on April 13, 1941, so as to enable Japan to focus on war with Britain and the United States. This pact, which matched the Molotov-Ribbentrop Pact of 1939 in allowing an Axis power to fight the Allies with a one-front war, reflected Hitler's assurances to Japan that he would attack Britain in 1941. This apparently provided Japan with an opportunity to take over British colonies. The alignment of Germany—the Soviet Union—Japan apparently provided the necessary opportunity. This helped explain Japanese anger when Hitler attacked the Soviet Union in June 1941.

In late 1941, pressure from Admiral Yamamoto Isoroku, the commander of the Japanese Combined Fleet, for an attack on the American naval base at Pearl Harbor in Hawaii (using the same military technique of the "knockout" blow as had been seen with Germany), as a prelude for covering Japanese invasions of Malaya, the Philippines, and the Dutch East Indies was successful. This plan won the day on November 3, 1941.[45] The devastating (but also unfinished) surprise attack on Pearl Harbor on December 7, "a date which will live in infamy" according to Roosevelt the next day, was to play a key role in American public memory,[46] as well as indicating that war did not begin with a declaration of hostilities, but, rather, with an event. In the event, most of the damaged warships returned to service and the battleships that were sunk were replaced with either more modern battleships or large aircraft carriers.

The Japanese suffered from the lack of a realistic war plan, a lack, already apparent in the case of Japan's attack on China, that was based on their assumption of their superiority over the Chinese and others. As with Hitler in the case of both Britain and the Soviet Union, a misleading conviction of the internal weakness of the opposing systems led the Japanese to a failure to judge resolve accurately, a failure that was important to the reading of the international system. In particular, there was a certainty that Britain and the United States lacked the willpower of Japan and Germany. The British and Americans were believed to be weakened by democracy and consumerism, an extrapolation of authoritarian views within Japanese politics and about

Japanese society onto other states. In the event, the initial Japanese ability to mount successful attacks, to gain great swathes of territory, and to establish an apparent stranglehold on the Far East and the Western Pacific behind a defensive perimeter, did not deter the Americans from the long-term effort of driving back, and destroying, Japanese power.

GERMANY AND THE UNITED STATES

Hitler followed the unexpected blow at Pearl Harbor by declaring war on the United States on December 11, 1941, as (ludicrously, but necessarily as an ally) did Mussolini. Hitler claimed that this declaration was in accordance with German obligations under the Tripartite Treaty with Italy and Japan signed on September 27, 1940. However, strictly, the terms of the treaty did not require such a cooperation, and Japan did not declare war on the Soviet Union in June 1941. Already angered by American cooperation with the British against German submarine (U-boat) operations in the Atlantic, operations that had already brought the war to the seas off North America, Hitler claimed that his decision to declare war was in response to American "provocations" in the Atlantic. Like Goebbels, Hitler saw the United States as part of a global Jewish conspiracy directed against Germany, and Roosevelt as the key instrument of this conspiracy. Already, on December 21, 1940, Germany had claimed that American assistance to Britain constituted "moral aggression."

In practice, there was no real appreciation in Germany of the impact of American entry and no sign of any informed analysis of the likely trajectory of war between Japan and the United States, nor of the consequences for Germany of war with America. This was an aspect of the more general failure of German Intelligence, and notably of Hitler's unwillingness to consider the views and capabilities of his opponents other than in terms of his own ideological suppositions, but also of a more general German politico-military naïveté.

As in 1917, when the United States declared war in large part because of the attack on its neutral rights represented by German submarine warfare,[47] there was a mistaken confidence in Germany that the U-boats would weaken the United States, providing Germany with a strategic weapon that could operate to strategic effect and negate the American effort in Europe. This was a belief necessary to Hitler's attempt to regain the initiative by a bold step. In reality, the U-boats only provided an operational capability, a situation exacerbated by serious mismanagement of the submarine fleet, notably a lack of large, long-distance boats.

In addition, this capability was gradually to be eroded by improved Allied anti-submarine warfare tactics and by the vast capacity provided by

American shipbuilding and, in particular, by the development of the latter thanks to large-scale investment, the availability of raw materials, and labor flexibility. By the end of the third quarter of 1943, the Allies had built more ships than had been sunk since the start of the war. The Americans could build them faster than the Germans could sink them.

Hitler, moreover, regarded the United States as weakened by deracination resulting from interbreeding, by consumerism, and by democracy, and as lacking in martial spirit, all views also shared by Japanese commentators. He felt that Japan and the U-boats would keep the United States busy until after Germany had successfully settled the war with the Soviet Union. In January 1942, Hitler told Lieutenant-General Hiroshi Oshima, the Japanese ambassador, that he did not know how he would defeat the United States, but such lucid moments were overtaken by an emphasis on will and by rambling fantasies.[48]

As a result of Hitler's declaration of war, the struggle was now truly global. The United States, in response, declared war on Germany on December 11, and its influence in Latin America was such that most of the world's remaining neutrals followed suit. Cuba, the Dominican Republic, Guatemala, Nicaragua, and Haiti also declared war that day. Honduras and El Salvador followed the next day, and Panama, Mexico, and Brazil in 1942. The declaration by Brazil, on August 22, was influenced by German attacks on Brazilian shipping as well as by American pressure and the Brazilian determination to exploit America's need for support.[49]

These entries marked a major blow to German diplomatic and espionage attempts to build up support in Latin America. In part, these attempts reflected the desire to exploit opportunities, not least those presented by local German populations and by authoritarian governments, such as the Peron dictatorship in Argentina. In part, the attempts were a product of the global aspirations of key elements in the German government. As in the case of Mexico in World War I, there was also a desire to weaken the United States by causing problems in its backyard. Thus, there was a strategic intention underlying Germany's Latin American policy. This policy had many flaws, not least encouraging American hostility and a central problem with implementation; the inability of Germany to give teeth to its hopes. This inability reflected British naval strength as well as the German focus on operations in Europe. Yet, despite the flaws of Germany's Latin American policy, there was a potential for causing trouble.

This potential was one of the victims of Hitler's decision to declare war on the United States. Instead, the Rio Conference in January 1942 saw the creation of the Inter-American Defense Board, which was designed to coordinate military matters throughout the Western hemisphere. In effect, the United States assumed responsibility for the protection of the region. It, not Germany, benefited from Britain's declining role. As a symptom of this declining role, in 1942, an Argentine navy transport raised the Argentine flag in the South Shetland island group contested with Britain and left a bronze plaque to record the occasion.[50]

America's success in Latin America, nevertheless, had limitations, which, in part, reflected the appeal of the German authoritarian model. Indeed, many Latin American states delayed entry into the war: Bolivia and Colombia until 1943, and Ecuador, Paraguay, Peru, Venezuela, Chile, and Argentina (in which there was much sympathy for Germany), until 1945. Although once they joined the war, none of the Latin American states played a major, let alone crucial, role in the conflict, their experience as combatants and neutrals reflected the global impact and nature of the struggle, at once military, political, ideological, and economic.

The state that played the leading military role was Brazil, which sent 25,000 troops to fight in the Italy campaign. As a result, Brazil was to be angered by postwar American sales of warships to its regional rival, Argentina, and refused to send troops to take part in the Korean War (1950–53). Northeast Brazil also played a staging role in the 1942 Operation Torch, an American amphibious attack on French-held North Africa. Mexico sent units to the Philippines in 1944.[51] Other late entrants into the war, all on the Allied side, were Liberia in 1944, and Saudi Arabia, Egypt, and Turkey in 1945.

Moreover, the alliance between Britain, the United States, and the Soviet Union led these three powers to declare war on those who were already at war with their allies. Thus, Britain and the United States went to war with Hitler's allies that had attacked the Soviet Union, while, keen to gain the spoils of victory, the latter went to war with Japan on August 8, 1945, after the end of the war with Europe. This was two days after the first atomic bomb was dropped on Hiroshima, an event that encouraged Stalin to act rapidly in order to seize opportunities.

CONCLUSIONS

World War II therefore was a number of conflicts, differently closely connected and partly separate. Their interactions were crucial to the diplomacy and strategy of the war. As with World War I, the causes of the conflicts varied greatly, while the number of these conflicts poses major problems for the statistical analysis of the causes of war. Again, as with the previous war, the context was crucial, notably the fear and opportunism created by the conflict, the extent to which it was not shut down but, instead, escalated dramatically, and the pressures from the two sides for new allies. The ideological drives of the war leaders and the regimes they controlled were central. They explained why alignments were both created and transformed. Ideological factors played a central role in the content and means of policies pursued, in short in both formulation and implementation.

NOTES

1. NA CAB 24/228 fol. 66.

2. B. Shillony, *Revolt in Japan: The Young Officers and the February 26, 1936 Incident* (Princeton, NJ: Princeton University Press, 1973); Sandra Wilson, *The Manchurian Crisis and Japanese Society, 1931–1933* (London: Routledge, 2002); Ian Nish, *Japanese Foreign Policy in the Interwar Period* (Westport, CT: Prager, 2002).

3. Louise Young, *Japan's Total Empire: Manchuria and the Culture of Wartime Imperialism* (Berkeley: University of California Press, 1998).

4. Richard Smethurst, *A Social Basis for Prewar Japanese Militarism: The Army and the Rural Community* (Berkeley: University of California Press, 1974).

5. Walter Skya, *Japan's Holy War: The Ideology of Radical Shintō Ultranationalism* (Durham, NC: Duke University Press, 2009).

6. Brian Victoria, *Zen at War* (New York, 1997; 2nd ed., Lanham, MD: Rowman & Littlefield, 2006).

7. Jeffrey Record, "The Use and Abuse of History: Munich, Vietnam and Iraq," *Survival* 49 (2007): 163–80; *Washington Post*, March 19, 2013.

8. Mikhail Saakashvili, former Prime Minister of Georgia, BBC Radio Four, 8:00 news, March 6, 2014; Daniel Johnson, "Reality Check for the West," *Standpoint* 61 (April 2014): 5.

9. Sally Marks, "Mistakes and Myths: The Allies, Germany, and the Versailles Treaty, 1918–1921," *Journal of Modern History* 85 (2013): 632–59.

10. Eugenia Kiesling, *Arming Against Hitler: France and the Limits of Military Planning* (Lawrence: University of Kansas Press, 1996).

11. Joe Maiolo, *Cry Havoc: The Arms Race and the Second World War, 1931–1941* (London: Penguin, 2010).

12. Adrian Gregory, *The Last Great War: British Society and the First World War* (Cambridge: Cambridge University Press, 2008).

13. Eire was then part of Britain and was represented in the Westminster Parliament, rather than being a Dominion.

14. Anthony Lentin, *Lloyd George and the Lost Peace: From Versailles to Hitler, 1919–1940* (Basingstoke: Palgrave, 2001), 103.

15. Keith Neilson, *Britain, Soviet Russia, and the Collapse of the Versailles Order, 1919–1939* (Cambridge: Cambridge University Press, 2006); Joe Maiolo, "Anglo-Soviet Naval Armaments Diplomacy Before the Second World War," *English Historical Review* 123 (2008): 352.

16. Hugh Ragsdale, *The Soviets, the Munich Crisis, and the Coming of World War II* (Cambridge: Cambridge University Press, 2004), 185.

17. Timothy Snyder, *Bloodlands: Europe Between Hitler and Stalin* (London: Yale University Press, 2010).

18. Richard Overy, *1939: Countdown to War* (London: Penguin, 2009).

19. Adam Tooze, *The Wages of Destruction: The Making and Breaking of the Nazi Economy* (London: Penguin, 2006), 665.

20. Talbot Imlay, *Facing the Second World War: Strategy, Politics, and Economics in Britain and France, 1938–40* (Oxford: Oxford University Press, 2003); David Edgerton, *Britain's War Machine: Weapons, Resources, and Experts in the Second World War* (Oxford: Oxford University Press, 2011).

21. Adam Claasen, "Blood and Iron, and 'der Geist des Atlantiks': Assessing Hitler's Decision to Invade Norway," *Journal of Strategic Studies* 20 (1997): 71–96.

22. Jeffery Gunsburg, "*La Grande Illusion*: Belgian and Dutch Strategy Facing Germany, 1919–May 1940 (Part I)," *Journal of Military History* 78 (2014): 101–58, esp. 141–54.

23. Churchill to Hoare, September 18, 1940, Churchill Papers, Cambridge, Churchill College.

24. Macgregor Knox, *Common Destiny: Dictatorship, Foreign Policy, and War in Fascist Italy and Nazi Germany* (Cambridge: Cambridge University Press, 2000); Robert Mallett, *Mussolini and the Origins of the Second World War, 1933–1940* (Basingstoke: Palgrave, 2003).

25. Galeazzo Ciano, *Diario 1937–39* (Milan: Enigma, 1980), 209.

26. Ciano, *Diario*, 140.

27. Hsiao-ting Lin, *Tibet and Nationalist China's Frontier* (Vancouver: University of British Columbia Press, 2006).

28. Richard Raack, *Stalin's Drive to the West: 1938–1945* (Stanford, CA: Stanford University Press, 1995); Evan Mawdsley, "Crossing the Rubicon: Soviet Plans for Offensive War in 1940–1941," *International History Review* 25 (2003): 818–65.

29. Oscar Pinkus, *The War Aims and Strategies of Adolf Hitler* (Jefferson, NC: McFarland, 2005).

30. Jeffrey Herf, *The Jewish Enemy: Nazi Propaganda during World War II and the Holocaust* (Cambridge, MA: Harvard University Press, 2006); Lorna Waddington, *Hitler's Crusade: Bolshevism and the Myth of the International Jewish Conspiracy* (London: Tauris, 2007).

31. Gabriel Gorodetsky, *Grand Delusion: Stalin and the German Invasion of Russia* (New Haven, CT: Yale University Press, 1999); David Murphy, *What Stalin Knew: The Enigma of Barbarossa* (New Haven, CT: Yale University Press, 2005).

32. Igor Lukes, "The Tukhahevsky Affair and President Edvard Benes: Solutions and Open Questions," *Diplomacy and Statecraft* 7 (1996): 505–29.

33. David Glantz, *Stumbling Colossus: The Red Army on the Eve of World War II* (Lawrence: University of Kansas Press, 1998).

34. Robert Citino, *Death of the Wehrmacht: The German Campaigns of 1942* (Lawrence: University of Kansas Press, 2007).

35. Paul Hanebrink, *In Defense of Christian Hungary: Religion, Nationalism and Anti-Semitism, 1890–1944* (Ithaca, NY: Cornell University Press, 2006).

36. Marius Turda, "In Pursuit of Great Hungary: Eugenic Ideas of Social and Biological Improvement, 1940–1941," *Journal of Modern History* 85 (2013): 588–91.

37. Dennis Deletant, *Hitler's Forgotten Ally: Ion Antonescu and His Regime, Romania 1940–1944* (Basingstoke: Palgrave, 2006).

38. James Herzog, "The Influence of the United States Navy in the Embargo of Oil to Japan, 1940–1941," *Pacific Historical Review* 35 (1966): 317–28.

39. Jay Taylor, *The Generalissimo: Chiang Kai-shek and the Struggle for Modern China* (Cambridge, MA: Harvard University Press, 2009), 172.

40. A. Cox, "The Effects of Attrition on National War Effort: the Japanese Experience in China, 1937–38," *Military Affairs* 32, no. 2 (October 1968): 57–62; E. Kinmonth, "The Mouse that Roared: Saitō Takao, Conservative Critic of Japan's 'Holy War' in China," *Journal of Japanese Studies* 25 (1999): 331–60.

41. Mark Peattie, Edward Drea, and Hans van de Ven, eds., *The Battle for China: Essays on the Military History of the Sino-Japanese War of 1937–1945* (Stanford, CA: Stanford University Press, 2010).

42. David Kaiser, *No End Save Victory: How FDR Led the Nation into War* (New York: Basic Books, 2014), 15.

43. Waldo Heinrichs, *Threshold of War: Franklin D. Roosevelt and American Entry into World War II* (Oxford: Oxford University Press, 1988); James Schneider, *Should America Go to War? The Debate Over Foreign Policy in Chicago, 1939–1941* (Chapel Hill: University of North Carolina Press, 1989).

44. D. F. Harrington, "A Careless Hope: American Air Power and Japan, 1941," *Pacific Historical Review* 43 (1979): 217–38.

45. Akira Iriye, *The Origins of the Second World War in Asia and the Pacific* (London: Longman, 1987); Donald Goldstein and Katherine Dillon, eds., *The Pearl Harbor Papers: Inside the Japanese Plans* (McLean, VA: Brasseys, 1993).

46. Emily Rosenberg, *A Date Which Will Live: Pearl Harbor in American Memory* (Durham, NC: Duke University Press, 2003).

47. Robert Tucker, *Woodrow Wilson and the Great War: Reconsidering America's Neutrality, 1914–1917* (Charlottesville: University of Virginia Press, 2007).

48. Ian Kershaw, *Fateful Choices: Ten Decisions that Changed the World, 1940–1941* (London: Penguin, 2007), 382–430.

49. Neill Lochery, *Brazil: The Fortunes of War: World War II and the Making of Modern Brazil* (New York: Basic Books, 2014).

50. Jon Wise, *The Role of the Royal Navy in South America, 1920–1970* (London: Bloomsbury, 2014), 112–13.

51. Frank McCann, *The Brazilian-American Alliance, 1937–1945* (Princeton, NJ: Princeton University Press, 1973); Michael Francis, *The Limits of Hegemony: United States Relations with Argentina and Chile during World War II* (Notre Dame, IN: University of Notre Dame Press, 1977); Robert Humphreys, *Latin America and the Second World War*, 2 vols. (London: Bloomsbury, 1981–82); S. I. Schwab, "The Role of the Mexican Expeditionary Air Force in World War II: Late, Limited, but Symbolically Significant," *Journal of Military History* 66 (2002): 1115–40.

Chapter Eight

The Age of the Cold War, 1946–89

A foreign bogey and the fear of foreign aggression must be held before them to stimulate their efforts for the new Five-Year Plan, and to persuade them that a large proportion of these efforts must be devoted not to improving the lot of the Soviet people. . . . The Soviet Union although confident in its ultimate strength, is nothing like so strong at present as the Western democratic world, and knows it.

—Frank Roberts, British diplomat in Moscow, 1946[1]

BREAKDOWN IN EUROPE

The Cold War that followed World War II was neither a formal nor a frontal conflict, but a period of sustained hostility involving a protracted arms race, as well as numerous proxy conflicts in which the major powers intervened in other struggles. In turn, the latter sustained attitudes of animosity among the major powers, exacerbated fears, and contributed to the determination to fund a high level of military preparedness. Moreover, these proxy conflicts could be large-scale, and were certainly not limited for those involved, for example in Korea, Algeria, Vietnam, Angola, and El Salvador.

Just as nineteenth-century theorists of international relations had focused on military conflict, so, notably in developing and applying realist theories, their Cold War successors concentrated on confrontation rather than conciliation. This concentration affected both political and military leaders, and, as a consequence, the public. Looked at differently, governments received the advice that they sought. For example, West German interest in détente in the late 1960s and early 1970s was not matched in the United States.

Wartime alliances frequently do not survive peace, and this was especially true of World War II because of the long-standing ideological division between the Soviet Union and the Western Allies. Moreover, this division was central to the polarization of politics that helped link World War II to the Cold War. There were also links in terms of the material and ideological resources the different parties were drawing on in both.

Arguably, the wartime alliance scarcely lasted for the war itself. Ideological and geopolitical tensions revived in its closing stages. By 1944, differences over the fate of Eastern Europe were readily apparent, especially over Poland, which Stalin was determined to dominate. Political rivalries within Eastern Europe, and between the allies, fed into the revival of the Cold War.[2] The latter had really begun in the aftermath of the Russian Revolution with large-scale foreign intervention against the revolutionaries in 1919 leading to a hot war. This failed to lead to their overthrow, although the independence of Estonia, Finland, Latvia, Lithuania, and Poland from Russia was secured. Moreover, a *"cordon sanitaire"* against Russian advance was put in place. Russian-backed subversion, directed in particular against British and French imperial interests, was an aspect of the subsequent Cold War of the 1920s and 1930s. It was to resume.

So also was conflict in Europe. Fighting broke out in Greece, where German evacuation in 1944 led to an accentuation of the conflict between left-wing and right-wing guerrilla groups seen across much of Europe, notably in Yugoslavia. In Greece, as in Yugoslavia, northern Italy, and China, there was a similar pattern: cruel occupation by Germany or Japan, a growing popularity (of sorts) of a Communist Party, and the outbreak of civil war between non-Communists and Communists, before, or once, the occupiers (Germans and Japanese) left or were expelled. Greece and Italy both were small, had semi-peninsular locations, and were confronted by powerful seaborne British or American forces that could bring their weight to bear. The situation was very different in Yugoslavia and China. In Greece, in order to thwart a left-wing takeover, conflict between Communists and Royalists was followed by military intervention by the British in support of the latter. Having arrived in October 1944, the British were fighting the Communist National Popular Liberation Army (ELAS) on behalf of the returned exile government two months later. The continuity between conflict during the world war and thereafter seen with World War I was also apparent with World War II.

After World War II ended, the Soviet Union initially lacked the atomic bomb, which had been successfully used by the Americans against Japan in August 1945. The previous month, Churchill, who was convinced of the value of atomic bombs, had reacted to the American atomic test by arguing that Soviet dominance could now be redressed.[3] However, helped by the startlingly fast demobilization of American combat forces in 1945–46 as soldiers returned to civilian society, a classic instance of the subordination of power considerations to popular demands, the Soviet army was apparently well placed to overrun Western Europe. It seemed that the Soviet army could only, in that event, be stopped by the West's use of nuclear weaponry, which therefore had to be the deterrent. In reality, had it advanced, the Red Army's logistical base probably would not have permitted it to have gone farther than the Rhine. The Red Army was renamed the Soviet Army in early 1946; Stalin wanted a more uplifting name than the Workers'-Peasants' Red Army to accord with the Soviet Union's new world status.

Nevertheless, once Harry Truman, who became president in 1945, revoked wartime Lend-Lease aid, the Soviets were no longer receiving spare parts for the backbone of their military motor transport, which was composed overwhelmingly of Ford and Studebaker trucks supplied by the Americans during the war. The equipment for routine tune-ups, such as spark plugs and distributors, and for oil changes, as well as other material, such as batteries, tires, inner tubes for tires, and oil and air filters, were no longer available. Nor were axles, drive shafts, gearboxes, and engine blocks. Tanks alone rolling ahead would not have been enough, and the horses sequestered during the war for Red Army transportation had to be returned home; Soviet agriculture was in very poor shape. Meanwhile, during any invasion, American and British bombers, escorted by their better fighter aircraft and better fighter pilots, would have been bombing the opposing army, front and rear. Their capabilities had been fully demonstrated in 1944–45 at the expense of German forces. Although poorly understood in the West, the Soviets were aware of their weaknesses. This weakness may well have deterred them, although it was not a deterrence comparable to that which might have been offered from the West, notably in the shape of American atomic bombs.

From 1945, despite serious military limitations on both sides, each side increasingly felt threatened, and seriously so, by the other, militarily and ideologically. This sense of threat helped define and strengthen the sides. The escalation of the conflict owed much to a clear sense of asymmetry in strengths, weaknesses, ideologies, and fears. Fear, indeed, was a driving element in the escalation of the Cold War, while also being a unifier of state and society, or, at least, a would-be unifier from the perspective of governments and commentators. This fear was to be seen across societies, including in popular culture.

Growing tension between, and within, states represented a failure of the hopes that it would be possible to use the United Nations to ensure a new peaceful world order. In 1943, the United States, the Soviet Union, Britain, and (Nationalist) China had agreed to the Moscow Declaration on General Security, including the establishment of "a general international organization, based on the principle of sovereign equality." The resulting United Nations was established in 1945, but it proved the setting, not the solution, for growing East-West tensions, and, although it could assemble forces, certainly did not lead to the world army that had been envisaged by some.

In addition, the Soviet victory over Nazi Germany and its allies had ensured a territorial and political settlement in which the Soviet Union affirmed its power. On January 10, 1945, Averell Harriman, the American ambassador in Moscow, reported (correctly) that, in occupied areas of Eastern Europe, the Soviet Union was trying to establish pliant regimes and that no distinction was drawn between former enemy states, such as Romania, and others, notably Poland, where, indeed, a conflict had begun between Soviet occupying forces and the non-Communist resistance to the Germans. He added that "the overriding consideration in Soviet foreign policy is the preoccupation with 'security,' as Moscow sees it. . . . The Soviet conception of 'security' does not appear cognizant of the similar needs or rights of other countries and of Russia's obligation to accept the restraints as well as the benefits of an international security system."[4]

This point was to be more generally true of Soviet attitudes and policy. Communism played a role in the equation. So did a sense of natural preeminence in what was regarded as "its region," a sense that is seen with Russia today. This attitude was also to be seen with Communist China (again as today) and, indeed, with the United States in Latin America. Thus, the combination of realist and idealist factors that determined the policies of states and, notably, empires, was apparent with the protagonists in the Cold War. American intervention, puppet regimes, and economic direction in Central America and the Caribbean were similar, up to a point, to Soviet activity in Eastern Europe. Albeit in a different ideological context, the concept of powers' regions or "spheres of influence" in nineteenth-century language, has recurred as a major one in the 2000s and, even more, 2010s, indeed has been central to the Chinese and Russian practice of international relations and bellicosity, which both poses a major challenge to the current norms of the world order and should also encourage a reconsideration of earlier examples.

Postwar, the Soviet Union retained all of the gains it had obtained in 1939–40 (from Poland, Romania, and Finland, as well as the Baltic Republics in their entirety). Thus, most of Lenin's losses in 1918–21 were reversed. The Soviet Union also added part of Czechoslovakia that had been annexed by Hungary in 1939, which was called Carpatho-Ukraine; as well as the northern part of the German province of East Prussia. The (southern) remainder of East Prussia went to Poland, which also gained substantial German territories on its western frontier: Silesia and Eastern Pomerania. Poland, in turn, lost more extensive territories, about 48 percent of prewar Poland, on its eastern frontier to the Soviet Union. This outcome, reversing the Polish gains in 1920–21, was rejected by the exiled Polish government. Now, as a result of the disintegration of the Soviet Union in 1991, these territories are parts of Belarus, Ukraine, and Lithuania, and, indeed, have helped cause differences and divisions within Ukraine. Poland, in effect, thus moved westward in

1945 as a country as a result of the postwar territorial settlement. Territories and cities were renamed accordingly.

Territorial changes were not on the scale of changes in 1918–23, but were still significant. Moreover, there was a large-scale movement of people. In 1945–46, nine million Germans fled or were driven west from the territories acquired by the Soviet Union, but, even more, from pre-1939 Czechoslovakia, Poland, and other countries. This flight was an aspect of the widespread displacement that the war had brought and that continued after its end. For example, Poles were moved out of the territories gained by the Soviet Union in the western Ukraine and were resettled in lands cleared of Germans. As a consequence of such movements, cities changed their ethnic identity. This displacement was part of the large-scale disruption that continued after World War II, notably, but not only, in Eastern Europe. Whereas the population movements within Europe in the 1950s and 1960s were primarily of labor (with families sometimes following), for example Italians to work in German factories, those in the 1940s were mostly of entire families as an aspect of what would later be termed "ethnic cleansing." Violence was involved in ethnic cleansing, but not genocide.

Thus, the end of World War II and the outset of the Cold War proved to have a pronounced demographic dimension that contributed greatly to the social strain, if not breakdown, associated with Soviet control and Communist dominance in Eastern Europe. However, as in Czechoslovakia and Poland, nationalists who were not Communist also supported these displacements. There was a determination to avoid the post–World War I situation of sizable ethnic minorities within the new nation-states. Japanese settlers were expelled from China (notably Manchuria) and Korea, or imprisoned for use as forced labor. At the local level, this was a form of war, as with the murderous expulsion of Italians from Istria in 1945 by Yugoslavs. Indeed, it was a war at the expense of the local population, one that is also seen very much in the modern world. Far less violently, the French drove the Italians from Corsica. The violent nature of the population moves exacerbated tensions well past the immediate postwar years. Dreams and fears of revenge and recovery contributed to Cold War politics among and about the defeated (West Germany, Japan), as well as the successful.

Ideology was an important factor in Soviet policy as, differently, in the American determination to create both a new world order and an Atlantic community, each described in terms of *the* (not *a*) "Free World."[5] However, it is easy to understand why the British diplomat Frank Roberts and some other commentators focused on geopolitics as well, not least as it enabled them to assert a degree of continuity in Russian/Soviet policy. There was this continuity but it also served the case of demonstrating the value of those who could explain it. Roberts, who was later to be influential in the British Foreign Office, reported from Moscow in March 1946:

> There is one fundamental factor affecting Soviet policy dating back to the small beginnings of the Muscovite state. This is the constant striving for security of a state with no natural frontiers and surrounded by enemies. In this all-important respect the rulers and people of Russia are united by a common fear, deeply rooted in Russian policy, which explains much of the high-handed behaviour of the Kremlin and many of the suspicions genuinely held there concerning the outside world . . . the fears aroused by foreign intervention after 1917 cannot yet have been eradicated from that of Western leaders, who, nevertheless, cooperated with the Soviet Union during the war.[6]

Concern about outside pressure was greatly accentuated by the war. For the Soviets, Eastern Europe served not only as an economic resource and an ideological bridgehead, but also as a strategic glacis. These attitudes can readily be seen in Putin's Russia, and serve as a reminder that geopolitical and strategic continuities draw on the same or similar assumptions. The element of fear was a key one. While claiming to represent the people, Soviet leaders feared them. This reflected the combination of their belief that the Communist Party was a cutting-edge elite and their concern that the people were malleable, and could therefore be misled by subversion.

The Soviet determination to dominate Eastern Europe was a key cause of tension, both local and international, not least because domination meant the imposition, through force and manipulation, of Communist governments. Seeking the finality of total control, the Soviets were not satisfied by the nuances of influence. The 1946 Polish elections, in which the Communists did well, were fraudulent. Force played an important role in the extension of Communist control. For example, King Michael of Romania abdicated on December 30, 1947, after the royal palace in the capital, Bucharest, was surrounded by troops of the Romanian division raised in the Soviet Union.[7]

The drive of Soviet policy appeared threatening to the other powers. The Soviets abandoned cooperation over occupied Germany, which, like Austria, had been divided into Soviet, American, British, and French occupation zones. They also imposed Communist governments in Eastern Europe, culminating with a coup in Prague in 1948. These moves vindicated Churchill's claim on March 5, 1946, in a speech given at Fulton, Missouri, in the United States, that an "Iron Curtain" was descending from the Baltic to the Adriatic. Truman had invited Churchill, from 1945 to 1951 head of the Conservative opposition in Britain, to visit his home state of Missouri and to give this speech. It was a response to Stalin's speech in Moscow on the occasion of electing delegates to the Supreme Soviet, given on February 9, 1946, in which Stalin spoke of the "inevitability of war with the West." This was a view he had long held. Nikita Khrushchev was to describe Stalin's idea of foreign policy as keeping the antiaircraft batteries round Moscow on twenty-four-hour alert.

1948, when Stalin also forcefully clamped down on what he called "nationalist-deviationist" Communist leaders, notably Josip Tito in Yugoslavia, was the bellwether year. This situation led to growing pressure for a Western reply. The coup in Prague in 1948 had a particular impact due to the rejection, in public discussion in the West, of the prewar Appeasement policy that had delivered Czechoslovakia to German control in 1938–39 and the view that this had simply encouraged renewed German expansionism. The democratic background of Czech politics was also relevant, as was Czechoslovakia's location as the most westerly of the countries that received a Communist government in the 1940s, and one that threatened the security of the American occupation zone in neighboring Bavaria. Thus, idealist and realist factors combined to make the Prague coup appear particularly sinister.

Soviet dominance of Eastern Europe was matched by an apparent threat to Western Europe. The extent of the threat is controversial, and there is debate about how serious it was, and how serious it was believed to be.[8] Nevertheless, there were scares, contributing to a culture of distrust.[9] Walter Bedell Smith, the American ambassador in Moscow, claimed, in September 1947, that the Kremlin felt it could safely create war scares as it knew that the United States did not want war and would not be the aggressor.[10]

Concerns in Western Europe led to the investigation of defense cooperation. There was interest in the idea of a Western European Third Force independent of the United States and the Soviet Union, and Britain and France accordingly signed the Treaty of Dunkirk in 1947. This agreement was directed against Soviet expansionism, although fear of a resurgent Germany also played a role, for much of the "lesson" of the 1930s focused on just such a resurgence.

However, in response to fears about Soviet plans, an American alliance, and commitment to Europe, appeared essential. In February 1947, the British, heavily indebted by World War II, acknowledged that they could no longer provide the military and economic aid deemed necessary to keep Greece and Turkey out of Communist hands. This was an urgent issue in light of the Greek Civil War, where the Soviet-backed Communists threatened to overthrow the Royalist government, and of strong Soviet pressure on neighboring Turkey. Instead, the British successfully sought American intervention, which was a key development in this stage of the Cold War and in altering the eventual contours of the struggle as a whole.

In contrast, American isolationism had contributed greatly to America not playing a comparable role in the earlier Cold War of the 1920s and 1930s. As a consequence, Americans tend to see the Cold War as beginning in the late 1940s. This is historically invalid, but captures the American attitude, and thus the key role not only of idealist factors but also of (American) realist approaches. To the Soviet Union, in contrast, there was far greater continuity from the Russian Civil War (in which the United States had played a role) to the post-1945 Cold War.

THE KOREAN WAR, 1950–53

Leading Western powers were tested not in Europe, but in distant Korea where there was a major war that was far greater in scale and consequence than being simply an Asian counterpoint to the Greek Civil War.[11] The Communists had overwhelmingly won in the Chinese Civil War of 1946–49, with the Nationalists retaining only Taiwan, but the Americans were determined that they should not be allowed further gains in East Asia. Sometimes a client kingdom of China, Korea had been conquered by Japan and become its colony in 1910. At the close of World War II, Korea, a hitherto united territory, had been partitioned: northern Korea had been occupied by Soviet forces and southern Korea by the Americans. In the context of the difficulties posed by Korean political divisions, and growing American-Soviet distrust, both of which sapped attempts to create a united Korea, they each, in 1948, established authoritarian regimes: under Syngman Rhee in South Korea and Kim Il-Sung in North Korea. There was no historical foundation for this division, each regime had supporters across Korea, and both wished to govern the entire peninsula.

The regime in North Korea, whose military buildup was helped by the Soviet Union, was convinced that its counterpart in the south was weak and could be overthrown, and was likely to be denied American support. The examples of China and Eastern Europe appeared to support such an analysis. The South Korean army, indeed, lacked military experience and adequate equipment, while the Korean Military Assistance Group provided by the United States was only five hundred strong. Moreover, in the late 1940s, with Communism triumphing in China and in Eastern Europe bar Greece, and waging struggles in Vietnam and the Philippines, the situation appeared propitious for expansion elsewhere. Foolish messages offered the Soviets encouragement. In December 1949, General Douglas MacArthur, American Commander in Chief Far East, told a visiting British journalist that the American defense line in the western Pacific specifically excluded South Korea and Taiwan (where the Chinese Nationalists had taken refuge in 1949), while specifically including Japan and the Philippines. The following month, Dean Acheson, the secretary of state from 1949 to 1953, reiterated a similar view at the National Press Club.

The bitter rivalry between the two Korean states, as each sought to destabilize the other, included, from 1948, guerrilla operations in South Korea supported by the Communist north. This rivalry led to full-scale conflict, on June 25, 1950, when the north launched a surprise invasion of South Korea.

It attacked with about 135,000 troops and using T-34 tanks and Yak airplanes provided by the Soviets that gave them an advantage over their lightly armed opponents. The South Korean army had no anti-tank weaponry, while the American infantry had the obsolete 2.5-inch bazooka, which was useless against T-34s.

As elsewhere in the Cold War, there was a significant linkage between international and local factors. In February 1950, Stalin had agreed to provide heavy guns to North Korea, in contrast to his position in March and September 1949 when he had rejected Kim Il-Sung's suggestion that an attack be mounted. In March–April 1950, Stalin moved further toward Kim Il-Sung's position, telling him that the Soviet explosion of an atomic device on August 29, 1949, and the treaty of alliance between China and the Soviet Union that had followed Mao's victory had made the situation more propitious. Thus, the Communist victory in China encouraged action against South Korea, in a variant of what was to be termed the domino theory. Conflict was continuous on the large scale as well as sequential and specific on the smaller scale.

Kim Il-Sung promised a quick victory in the "Fatherland Liberation War." Stalin made agreement, however, conditional on Chinese support and said that, if the Americans intervened, he would not send troops, and thus slacken his focus on Europe. This approach increased North Korea's reliance on Chinese backing while also committing China. Given Stalin's position, Mao agreed, although he needed to consolidate his position in China and would have preferred an emphasis on gaining Taiwan rather than invading South Korea. Indeed, the Korean War helped save Taiwan from a Communist takeover, both by postponing a possible Chinese attack and, once distrust of Mao escalated, by greatly increasing America's willingness to help Taiwan.

In June 1950, the South Koreans were pushed back by the invading North Koreans, the capital, Seoul, falling on June 28. However, enough units fought sufficiently well in their delaying actions during their retreat south to give time for the arrival of American troops. This element of opportunity was highly important. As with the later Vietnam War, this was a conflict in which America's local ally played a key role. American forces entered combat in Korea from June 30 as the North Korean invasion led to intervention by an American-led UN coalition. This wide-ranging coalition was determined to implement and maintain policies of collective security and containment, and was concerned that a successful invasion of South Korea would be followed by Communist pressure elsewhere, possibly on West Berlin or on Taiwan, but equally, at any other point that might appear opportune. After the South Koreans, the leading UN contingent was American, and the second largest was British. Among the large number of international participants, the Canadians, French, and Turks were prominent. At the same time, underlining the specific character of the causes of war, some American allies did not participate. Angered by American arms supplies to Argentina, Brazil, unlike in World War II, refused to send troops.

The Americans also provided most of the air and naval power, as well as the commander, MacArthur. The UN forces benefited from the backing of a relatively stable South Korean civilian government and from a unified command: MacArthur's position as Supreme Commander of UN forces in Korea ensured control over all military forces, including the South Korean army, and provided a coherence that was to be lacking in the Vietnam War. American capability was enhanced by the presence of their occupation forces in nearby Japan, by American air bases there, and by the logistical infrastructure and support services provided by Japanese facilities and resources including harbors.

Allied assistance was important in restricting criticism within the United States of the role of other powers. On August 2, 1950, the Chancery at the British Embassy at Washington, forwarding a report by Strafford Barff, the Director of British Information Services in Chicago, noted that "the cynical comments about America being left alone to do the fighting are mainly confined to the light-weight and lunatic newspapers and have not been indulged in by the more responsible press which understands the implications on a world-wide basis of Soviet strategy." Barff on July 31 reported both press comment that European countries, including Britain, were psychologically unfit to wage war and that "an increasing number of people believe that Korea is the prelude to World War III and that it will follow soon."[12]

The issue of psychological fitness to wage war was important to discussion during the Cold War, as also with the previous world wars. The issue reflected the role of conscription. The equations of opportunity and fear that were so important to the causes of war in part rested on assumptions about this fitness, assumptions that often rested on glib views about the weakening effects of factors such as consumerism or Communist rule.

Thanks to their major role in World War II, the Americans were better able than they would have been in the 1930s to fight in Korea. Nevertheless, since 1945, due to postwar demobilization as the "peace dividend" was taken, there had been a dramatic decline of available manpower and matériel, and they were scarcely driven by a large and well-prepared military to opt for war. The number of amphibious ships had fallen from 610 in 1945 to 81 in 1950, there was a grave shortage of artillery units, and, in 1949, the American army only contained one armored division. American fighting effectiveness had also declined, as was shown by the experience of some American units in the first year of the Korean War. The American occupation forces sent from Japan to South Korea in late 1950 were out of shape and performed poorly in July. Many of the National Guard units dispatched were inadequately trained and equipped. Barff had noted on July 31, "The inadequacy of American arms and reported inefficiency of some officers and men have come as a great shock."[13]

The Americans were almost driven into the sea at the end of the peninsula in the first North Korean onslaught, but, once reinforced, managed, fighting well, to hold the Pusan perimeter there against attack. The situation was rescued by Operation Chromite, a daring and unrehearsed landing in very difficult tidal conditions on the Korean west coast at Inchon on September 15, 1950. This landing applied American force at a decisive point. Carried out far behind the front, and with very limited information about the conditions, physical and military, that they would encounter, about 83,000 troops were successfully landed. They pressed on to capture nearby Seoul.

This success wrecked both the coherence of North Korean forces and their supply system, which had anyway been put under great strain by the advance toward Pusan, and also achieved a major American psychological victory that was not to be matched in the Vietnam War. The capture of Seoul enabled the American forces in the Pusan area to move north. The North Koreans were driven back into their own half of the peninsula, which changed the nature of the war and, arguably, caused a new one: "defense" is not the same as "forward defense," let alone a conflict to overthrow the North Korean regime. On October 7, 1950, American forces crossed the 38° North parallel dividing North and South Korea. They moved north toward the Chinese frontier, advancing across a broad front against only limited resistance.

The UN advance was not welcome to the Chinese. In effect, in a new war, albeit an undeclared war, they suddenly intervened in October, exploiting the overconfidence of the Americans, notably, but not only, MacArthur. From July, the Chinese appear to have begun preparing for intervention, and certainly built up large forces near the border. Communist success in the Chinese Civil War (1946–49) had encouraged Mao to believe that technological advantages, especially in airpower, which the Americans dominated, could be countered, not least by determination. This was a "realist" belief in the prospect of victory, albeit one heavily conditioned by ideological suppositions. However, as also by the Japanese in World War II, American (and, more generally, Allied) resilience, resources, and fighting quality were under-estimated by the Chinese, in this the sole war between any of the world's leading military powers since 1945, so far. Mao felt that UN support for Korean unification threatened China and might lead to a Nationalist (Guomindang) *revanche*, saw American support for Taiwan as provocative, and was also keen to present China as a major international force. In turn, Stalin, to whom Mao appealed for help, wanted to see China committed against the United States. This was the price for meeting Chinese requirements for assistance with military modernization. Stalin promised help in the event of the Americans invading China as a result.

It was not only the Western powers that faced problems in adjusting to the major changes in the international system. Indeed, under Stalin, there was a reluctance to see other Communist powers as more than clients, and notably so in the case of Asian states. Mao's China was a key recipient of this condescension and attempted direction. Thus, the Soviet Union was reluctant to return to China the major warm-water Manchurian ports of Port Arthur and Dalian that it had taken from Japan in 1945, and, despite Mao wanting the ports returned, did not do so while Stalin remained alive. This stance represented an important continuation from pre-Communist Russian interest in Manchuria: Port Arthur had been used as a naval base then until captured by Japan in 1905.

Chinese intervention had not been anticipated by MacArthur, who had believed that, by maintaining the pace of the offensive and advancing to the Korean-Chinese frontier at the Yalu River, he would end the war. This advance had been authorized by the Joint Chiefs of Staff, and there was encouragement from CIA reports that direct intervention by China and the Soviet Union was unlikely. Despite a Chinese warning on October 3, 1950, via the Indian envoy in Beijing, of action if the UN forces advanced into North Korea, it was believed that the Communist leadership was intent on strengthening its position within China and that China lacked the resources for intervention abroad. MacArthur ignored the more cautious approach taken by Truman and his secretary of state, Dean Acheson. Concerned about the Chinese response, Truman instructed MacArthur to use only South Korean forces close to the Chinese border, but MacArthur was insistent that American troops be employed. MacArthur told Truman that it was too late in the year for the Chinese to act in strength, and, after they initially intervened from October 19 in fairly small numbers, he neglected evidence of Chinese troops in Korea, not least of Chinese prisoners. The operational success MacArthur had shown in the Inchon operation was not matched by adequate strategic assessment on his part.

Although his hubris was partly responsible, there were also serious weaknesses in American command and control reflecting the improvised way in which the conflict was being fought. In addition, the belief that airpower could isolate the battlefield, and dominate larger ground forces there, led to misplaced confidence. These points throw suggestive light on the extent to which the theory, doctrine, and infrastructure of nuclear deterrence might later have not worked had there been conflict between the major powers. The serious miscalculations shown in Korea in 1950, by the Americans, and also earlier by Communist planners who did not anticipate an American intervention in South Korea, did not provide encouraging instances of the effectiveness of deterrence. As with other conflicts, the Korean War showed how what was intended as a distinct and very limited conflict became a larger-scale and, for a while, indeterminate one.

NUCLEAR WAR

The nuclear confrontation between the United States and the Soviet Union established a new context for discussing international relations and deterrence.[14] The difficulty of working out new rules was compounded by mutual misunderstanding and distrust, and by the pace of particular crises. This was the background to much of the writing about international relations (IR) theory. Indeed, it exemplified the point that both history and IR studies were affected by changing contexts and values, such that IR theory was in practice very much historicized. Academics grew up taking part in nuclear air raid drills.[15] In the context of nuclear confrontation, deterrence theory and discussion of escalation both came to the fore.[16]

In practical terms, this process was greatly encouraged by the Cuban Missile Crisis of 1962, a "near-war," in which geopolitical competition and strategic anxieties focused on the risks of an American invasion of Cuba and of the Soviet deployment on Cuba of weapons capable of delivering atomic warheads. The Soviet attempt covertly to base missiles on Cuba and then to "present" them to the world failed dramatically when the Americans obtained visual proof of the missiles on Cuba with pictures taken by U-2 aircraft on October 14. The situation on Cuba then got seriously out of the grasp of the Soviet leadership, with no direct control from Moscow over the already-deployed and ready, nuclear-armed, short-range missiles that could be used against a potential American invasion; with no direct way of communicating with the leadership in Washington; and with only very limited control over a very bellicose Fidel Castro.

President John F. Kennedy reacted with a naval blockade that some in his government considered an act of war. This blockade, which was implemented very fast with allied help, was a victory for the significance of conventional naval power in the nuclear age. The Soviet view of Kennedy as a figure with whom they could negotiate helped ensure an informal solution to the crisis, albeit without a formal treaty.

The crisis demonstrated the danger of a nuclear confrontation, which encouraged the creation of systems for limiting future crises, notably the "hot line" between Moscow and Washington. In 1972, following an American proposal in 1968 and talks beginning in 1971, INCSEA, an American-Soviet Incidents at Sea agreement, helped prevent unwanted "accidents" between the two navies and such incidents spinning out of control. This was a classic example of the use of creative diplomacy to help lessen the causes of crisis. The agreement stipulated a range of measures including steps to avoid collisions, care in maneuvering, not stimulating attacks, and informing vessels when submarines were exercising near them.[17]

THE VIETNAM WARS

Alongside a nuclear confrontation that threatened total war on an unprecedented scale, the Cold War involved limited warfare, of which the lengthiest
for the United States was the Vietnam War. This conflict exemplifies the
extent to which the causes of war are very different for the combatants,
including those on the same side. Moreover, the course of that war demonstrates the process by which the "causes" of war can vary during its course as
new pressures come into play. Such a remark may appear counterintuitive; a
war might seem to have a set of causes that relate to its origins. That is
indeed the case, but the course of a conflict also involves fresh opportunities
to change the effort or vary the goals. As a result, "causes" should be seen in
a dynamic fashion.

Vietnam, part of the French empire as a result of nineteenth-century expansionism, witnessed a bitter insurrectionary struggle after World War II.
The French failure to suppress opposition led to withdrawal in 1954 with the
north, which was left under a Communist government, and the south, which
was left under an anti-Communist one. In turn, the American-supported
government in South Vietnam faced a Communist rebellion by the Viet
Cong, which led to more overt American intervention. In turn, in a process
that had begun before American intervention, a process encouraged by China, forces from North Vietnam moved south to help the Viet Cong. By 1963,
there were 16,000 American military advisers, but the South Vietnamese
army, which was wracked by internal (including religious) differences, was
not in command of the situation. In part, this was because it was having to
respond to its opponents and had a large area to defend, but there was also a
command culture focused on caution and firepower, and that could not grasp
the dynamic of events. Being on the defensive meant that its opponents were
able to dictate the pace of campaigning, and thus force on them a situation of
a new set of potential causes of renewed or expanded conflict.

By 1965, in the face of the North Vietnamese and Viet Cong moving, as
they thought, into the last phase of Mao Zedong's theory of revolutionary
war, and accordingly deploying large forces operating in concentrated units,
the South Vietnamese army was on the verge of collapse. This situation led
to a major increase in American commitment in order to preserve the cred-
ibility of American power, an element frequently important to the causes of
war, and to force war on the Communists in an area where the Americans
could intervene. This was not the case for China in the late 1940s.

Thus, to President Lyndon B. Johnson (r. 1963–69), the war was a necessary demonstration of resolve, a strategic goal that rather swallowed the
specific problems of winning success in South Vietnam. This was a sequel to
what was seen as the failure, both political and military, to prevent Cuba
going Communist, an assessment that exaggerated what the United States

could readily have achieved once Fidel Castro was established in 1959. American policy makers were affected by the "domino effect," a belief that failure in Vietnam would be followed by a sequential Communist advance into neighboring areas, first Laos, Cambodia, and Thailand, and then Malaysia. This assumption, that of averting imagined consequences, has frequently been very important in the causes and course of conflict.

Denying the American success was presented by the North Vietnamese as a way to bring victory on the grounds that American willpower to sustain the struggle was less than that of their opponents. However, although easy to claim, that did not suffice for the North Vietnamese and, more particularly, ensured that they found it difficult to shape the conflict. This position was accentuated by American claims, based on misleading statistical indices of victory, that the war was going well. This situation helped ensure the launch of the Tet Offensive, a major offensive by the North Vietnamese and Viet Cong in 1968 designed to show to the American public that their army was failing, and also to demonstrate to the South Vietnamese that this army could not protect them.

Having defeated these attacks, American effectiveness in counterinsurgency increased from 1968, but, in part for tactical and operational reasons, it still proved difficult to "fix" opponents and to force them to fight on American terms, and thus to "cause" a war accordingly. Nevertheless, in 1969, the Americans inflicted serious blows on the Viet Cong, who had lost many of their more experienced troops in the Tet Offensive. The Viet Cong achieved little in 1969, and their attacks suffered heavy losses.

Yet, although the Americans were able to repulse attacks, their counterinsurgency strategy was hit by the unpopularity of the South Vietnamese government, by Viet Cong opposition and intimidation, and by growing vocal domestic American criticism of what appeared an increasingly intractable conflict. The last encouraged the Americans to shift more of the burden back on the South Vietnamese army, in short to change the war. This army had some good commanders and units but was not up to American expectations. Thus, prefiguring the situation in Afghanistan after the Soviet withdrawal in 1989, although the Viet Cong and North Vietnamese did not win in the field in 1968–72, they benefited greatly from shifts in the military and political contexts. This element in victory throws considerable light on the issues that should be addressed when evaluating the decisions that led and lead to war.

At the strategic level, these shifts in context included growing pressure on American interests elsewhere, and notably so as a result of Soviet support for Arab rearmament and intransigence after the Six-Day War of 1967 between Israel and its Arab neighbors. The Soviet deployment of more warships in the Mediterranean increased the pressure on the United States. There was also concern about the situation in Korea. Given the American determination to avoid World War III, a determination shared by Soviet policy makers, these pressures were important. Thus, war or its prospect in one area affected war or its prospect in another, a conclusion that remains valid today.

The balance of failure in Vietnam, that of failure by both sides, a point true in many conflicts, but one that often leads to pressure for renewed efforts, continued. North Vietnam demonstrated this balance in 1972, in one of the major offensives of the period. The casualties inflicted on the Viet Cong in, and after, the Tet offensive, as well as the inability of American air attacks to destroy North Vietnam's war-supporting capability and logistical system, had caused a new war or stage in the war. This was characterized by a greater reliance on North Vietnamese forces rather than on the Viet Cong, and the possibility for the use of conventional forces in a conventional Soviet-style operation. In March 1972, the North Vietnamese launched the Nguyen Hue campaign (or Easter Offensive) across the Demilitarized Zone between North and South Vietnam.

The failure of the Nguyen Hue campaign meant that the North Vietnamese would need to follow the route of negotiation in order to move forward in the Vietnam War. This course was encouraged by the 1972 American rapprochement with China, a step of great strategic significance that, like the earlier overthrow of the left-wing nationalist government in Indonesia in 1965–66 by the military, encouraged by the CIA, made it less serious for the Americans to abandon South Vietnam. Indeed, as a result, there was a strategic "victory" of a sort for the United States in the Vietnam War. The Paris Peace Agreements of January 1973, during the negotiation of which in December the Americans threatened to use nuclear weapons, were followed by American withdrawal two months later. On the eve of the American withdrawal in 1973, neither side had won the war on the ground, a repetition of the situation for the French there in 1954 and in Algeria in 1962, which was not a comparison the Americans would have welcomed. However, the Americans, like the French in 1954, were under serious fiscal pressure and suffering from rising domestic problems.

The conflict continued, as a new war caused by the continuation of the previous one, now with the two Vietnams the combatants. In April 1975, South Vietnam was overrun, in the Ho Chi Minh campaign, by a renewed invasion from the North. The withdrawal of American forces and the total fall of South Vietnam were not the limits of conflict in the region. In 1975, the Communists also overthrew their opponents in Cambodia and Laos.

That could be the end of this section, but to do so would be to make the mistake of isolating the American intervention and its causes, and failing to work through the context and the consequences. There was a major falling out in the regional dimensions of the Communist bloc, one that stemmed from the Sino-Soviet split. Vietnam, which looked to the Soviet Union, in 1978–79 conquered Cambodia, which looked to China. In response, believing that Vietnam ought to be taught a lesson and fearing a fundamental

Soviet threat to Chinese security, in February–March 1979, the Chinese attacked Vietnam with 500,000 troops, inflicting much devastation, only to find that greater Vietnamese guerrilla warfare experience, combined, on the part of the Chinese, with poor logistics, inadequate equipment, and failures in command and control, led them to withdraw, and without forcing the Vietnamese forces to leave Cambodia. Although far larger in scale and longer than its attack on India in 1962, this was also a more limited war than the Chinese intervention in Korea in 1950–53, notably because China was not fighting a major power. It proved far easier for China to restrict its commitment in Vietnam than in the case of the United States in Vietnam or the Soviet Union in Afghanistan. The Chinese-Vietnamese war has largely been "lost" to attention.

Low-level conflict continued in Cambodia, with China backing rebels after its protégé, the Pol Pot government, was overthrown by a Vietnamese invasion in 1978–79. Moreover, border conflict continued between China and Vietnam until 1991, with much of it on a large scale and very costly in lives.

Nevertheless, there was no major conflict in East or Southeast Asia after the 1970s. Partly as a result, the capacity of the region to lead to conflict was underplayed until the situation dramatically changed in the mid-2010s. Indeed, the focus on the Islamic world in the meanwhile was a product not only of the inherent importance of warfare there, but also of a relative significance arising from a lack of conflict in East and Southeast Asia, an area of far greater and rapidly growing economic weight. China's economic rise in the 1970s–2000s was very much achieved through integration with the American-dominated global system and without conflict or even confrontation. The situation did not alter until the 2010s, when tensions rapidly escalated.

Another approach would be to ask how best to define a major war. The Vietnamese invasion of Cambodia in 1978–79 involved 150,000 troops and followed conflict between Cambodia and Vietnam in 1975–78. This 1978–79 invasion was initially resisted by conventional means, leading to the loss of about half of the Cambodian army, until recourse was had to guerrilla operations from bases in Thailand. This makes the dating of the war more problematic. Subsequently, the Vietnamese retained a large force in Cambodia, indeed 180,000 of their 1.26-million army in 1984, a year in which major efforts were launched against the guerrillas. In 1989, the Vietnamese withdrew their forces, a change that was linked to a cut in the Vietnamese army by about half. About 15,000 Vietnamese troops had been killed during the occupation. Peace came in 1999 when the Khmer Rouge, the key resistance element, no longer enjoying Chinese support, was completely dissolved.

A conflict on this scale would have been regarded as major elsewhere in the world. In Southeast and East Asia, a conflict, with embedded reporters, in which the United States was to the fore had been replaced by a situation in which regional powers, and notably the regional superpower, China, were the key players, while the American role was essentially offshore. That is a situation that has continued to the present.

At the same time as American intervention in the Vietnam War, there were a number of other wars between states, most prominently in the Middle East and in South Asia. In part, these conflicts reflected disputes left over from the period of colonial rule, but ideological divisions were more to the fore, notably religion, both in the Middle East, with conflict between Israel and its Muslim neighbors, and between Hindu-dominated India and Muslim Pakistan. Great power alignments were also significant, particularly in providing armaments, other forms of support, and encouragement. Moreover, regional powers could intervene, as Libya did in Chad in the 1980s. There could also be continuity with precolonial conflicts as with the Biafran war in Nigeria in 1967–70, a struggle in which ethnic and religious difference was focused on a separatist struggle: that by the Ibo in Biafra, the state established in southeast Nigeria. The same was also true of conflict between north and south in Sudan. Such conflicts had many elements in common with state-to-state warfare, but were also harder to alleviate, let alone settle, due to the fixity of territorial borders.

AFGHANISTAN

The Soviets, who had sought to woo Afghanistan as an ally against Britain prior to World War II, had been major aid donors to Afghanistan from the 1950s, taking its side in a frontier dispute with American-backed Pakistan. This role matched the earlier, nineteenth-century, geopolitical tension between British India and Russian intervention in Afghanistan, again indicating the continuation of earlier alignments, albeit in a different context and with policies that reflected ideological norms. In 1973, the Afghan monarchy was overthrown in a coup and an authoritarian strongman, Mohammed Daoud Khan, a cousin and the brother-in-law of King Zahir Shah (r. 1933–73), who was also a former prime minister, took power. Backed by a group of Soviet-trained officers, Khan was willing to accommodate the Soviet Union. The coup was seen as an extension of Soviet influence, which certainly increased; but Khan was a nationalist, not a Communist. In turn, in the Saur Revolution on April 27–28, 1978, Khan was overthrown and killed in a coup mounted by the Soviet-backed People's Democratic Party of Afghanistan. The presidential palace fell to a tank assault assisted by air strikes.

Bitterly divided between the Khalq (Masses) and Parcham (Banner) factions, the new government responded to opposition with repression. Its attempts to reform a largely conservative Islamic society, particularly with land reform and equality for women, both staples of the Communist prospectus for supposedly backward societies, led to rebellions from late 1978. The government met these with considerable brutality, including the bombing of recalcitrant cities, notably Herat. After a coup from within the regime, on September 16, 1979, did nothing to stem the tide of chaos, the Soviets militarily intervened in Afghanistan from December 25, 1979, violently overthrowing the government of Hafizullah Amin (who was killed), and installing Babrak Karmal as a client president.

The Soviet intervention, which was not intended to launch a war, still less the bitter, intractable, and lengthy one that resulted, appears to have resulted from a number of factors; governments are not obliged to provide academics with a clear, or politics-free, explanation. Concern about the stability of their position in neighboring Central Asia was a factor, notably anxiety that Afghanistan's problems could spill over into Soviet Tajikistan; until the breakup of the Soviet Union in 1991–92, it was a Central Asian power. There was also an unwillingness to see a client state collapse. Moreover, there was anxiety that the Amin government might turn to China, thus extending the threat to Soviet borderlands. Chinese delegations were traveling to Afghanistan in 1978.

Contemporary Western suggestions that the Soviet Union was seeking to advance to the Indian Ocean and the Persian Gulf appear overstated, although the Soviets were interested in developing a major air base of Shindand near Kandahar, from which the entrance to the Gulf could be rapidly overflown. The takeover of Afghanistan offered a way to follow up the eviction of the Americans from neighboring Iran in 1978–79 and also to put pressure on neighboring Pakistan, which was seen as a Chinese ally and as an enemy of India, the Soviet Union's largest South Asian friend. There was also an opportunity to hit American prestige and thus its ability to maintain alliances and elicit support.

In a process of escalation, the Soviet intervention in Afghanistan helped lead not only to Western alarmism, but also to pressure for an active response, both there and elsewhere. The Soviet invasion was regarded not as a frontier policing operation designed to ensure a pliant government, but as an act of aggression that needed to be countered. This view drew on a pronounced tendency, seen throughout the Cold War, to exaggerate Soviet political ambitions and military capability—a tendency matched by Soviet views of the Americans. It was, however, difficult for the Americans to acquire accurate information. The Soviet Union was a closed society with little in the way of open information or of possibilities for Western journalists and politicians to talk to members of the Central Committee or even the Politburo. Nothing of substance was offered to the American senatorial delegations that visited Moscow. In his State of the Union address to Congress in January

1980, President Jimmy Carter (r. 1977–81) warned that the Afghan invasion "could pose the most serious threat to peace since the Second World War."

REAGAN AND COLD WAR ESCALATION

A determined opponent of Communism, as well as an improviser at the level of implementation, President Ronald Reagan, the victor of the 1980 election, was happy to be associated with a marked intensification of the Cold War. However, Carter had already adopted a moral approach to the Soviet Union that differed from the realpolitik of President Richard Nixon (r. 1969–73) and that, from 1977, took him increasingly away from détente. Carter's stance owed much to concern about harsh Soviet human rights policies. This stance was also a product of the Soviet modernization of their nuclear missiles, specifically the deployment of the SS-20 missile, as well as of opposition to Soviet expansionism in the Third World. Mobile, accurate, and armed with nuclear warheads, the SS-20s were designed to be used in conjunction with conventional forces in an invasion of Western Europe. Carter's criticism angered the Soviet leadership, leading them not to negotiate seriously on a number of contentious issues, which provided Reagan with a further rationale for his anti-Soviet attitudes.[18] Although he backed down in 1978, in the face of a Soviet propaganda blitz, from deploying the enhanced radiation neutron bomb, Carter began the military buildup that Reagan continued and for which he received at the time most of the credit. Moreover, Carter adopted a more active regional stance after the Soviet invasion of Afghanistan, for example pressing Somalia to provide access to the Indian Ocean port of Berbera.

Ironically, although Reagan regarded Carter as hopelessly idealistic, he was equally ideologically motivated in foreign policy and, indeed, rejected realpolitik, a trajectory that was also to be seen with President George W. Bush (r. 2001–9). Reagan's earlier political career, notably his right-wing stance in the 1960s, in his successful 1966 campaign to be governor of California and in his unsuccessful campaigns for the Republican nomination in 1968 and 1976, had made his views clear. Reagan restated these views during the election campaign, declaring in June 1980 that "the Soviet Union underlies all the unrest that is going on." Reagan was encouraged in his anti-Communist resolve by Margaret Thatcher, Britain's resolute Conservative prime minister from 1979 to 1990, who had seen Carter as insufficiently firm, and by Karol Wojtyla, John Paul II, pope from 1978 to 2005. However, there was no comparable drive from elsewhere in Western Europe.

Relations with Thatcher were eased by American logistical support and diplomatic forbearance during Britain's 1982 war with Argentina over the Falkland Islands. There was no repetition of the serious undermining of

Britain seen in the 1956 Suez Crisis. At the same time, the possibility that key NATO anti-submarine naval assets would be sunk in that conflict served as a reminder of the dependence of the Cold War on other agendas.

Reagan was not prepared to accept that the Cold War should, or could, end in a draw enforced by a threatening nuclear peace and the related arithmetic of deterrence. Instead, Reagan, Thatcher, and John Paul II were determined to defeat what they each saw as an immoral and dangerous ideology. Reagan referred to the Soviet Union as an "evil empire" in a speech on March 8, 1983. This remark provoked a very hostile media reaction. This was a theological as much as a political judgment of foreign policy. At the same time, Reagan's remark was mild compared with the gross cartoons and inflammatory and inaccurate articles that were a staple of the Soviet newspaper *Pravda*, a point that was rarely made.

Already on September 2, 1981, Reagan had warned that the United States was prepared to pursue an accelerated nuclear arms race with the Soviet Union. Visiting Britain in June 1982, Reagan addressed British parliamentarians in the Royal Gallery in the Palace of Westminster, calling for a "crusade for freedom" and for discarding Marxist-Leninism on the "ash-heap of history." Europe was the center of his concern: "From Stettin on the Baltic to Varna on the Black Sea," there had been no free elections for three decades, while Poland, where the independent trade union movement Solidarity was under assault was, he declared, "at the center of European civilization." This approach did not accept the idea that Poland should be securely located in the Communist bloc. Reagan also provided encouragement and support for the Afghan resistance to Soviet occupation, notably, from 1985–86, shoulder-fired ground-to-air missiles. This support strengthened the resistance, as well as providing the Soviets with a factor to blame when explaining continued opposition there.

There has, however, been considerable controversy over whether Reagan had a grand strategy for confronting and weakening Communism, as argued, for example, by John Lewis Gaddis.[19] Alternatively, it has been claimed that there was no such strategy but, rather, a set of beliefs, notably the clashing aspirations of destroying Communism, and ending the risk of war: a "crusade for freedom" alongside "peace through strength."[20]

Far from being cowed by Soviet military developments and deployments in the 1980s, not least the creation of a major naval capability and the deployment of intermediate-range missiles, the American government and military responded with higher expenditure and a vigorous determination to develop doctrines that would enable an aggressive response, on land, sea, and air, to any Soviet attack. The former brought profit to the military-industrial complex, which, in turn, helped ensure that particular localities, and thus politicians, had an incentive to support military expenditure.[21] This buildup was accomplished without conscription: only so much popular support was required.

On the part of the American military, there was a focus on how war with the Soviet Union could be won without a massive nuclear exchange, and therefore how such a war could be waged. A space was to be created for the causes of war, including that of responding to Soviet attack. The example of Israeli success in the Yom Kippur War of 1973, and the doctrine of AirLand Battle, led to a stress on the integration of firepower with mobility in order to thwart the Soviet concept of Deep Battle. Proposing an effective synergy between land and air, and an intermediate level between the tactical and the strategic, this doctrine was designed to permit the engagement and destruction of the second and third echelon Warsaw Pact forces, at the same time that the main ground battle was taking place along the front, which suggested that NATO would be better placed than had been argued to repel a Soviet conventional attack in Europe. This doctrine led to an emphasis on the modernization of the conventional army. New weapons systems included the Blackhawk helicopter introduced in 1979; the M1A1 Abrams tank, deployed from 1980; the Bradley Fighting Vehicle, designed to carry a squad of infantry and armed with a TOW (tube-launched, optically tracked, wire command data link) missile system, introduced in 1981; and the Apache attack helicopter, equipped with radar and Hellfire missiles, introduced in 1986.[22]

Looked at more critically, the Americans were over-reliant on the capabilities of what airpower could achieve. This was also seen at sea. In 1982, in the Northern Wedding naval exercise, American carrier battle groups approached close enough to the Kola Peninsula to be able to launch planes carrying a full load to attack the Severomorsk naval base and then return. Successful war therefore seemed to be a possibility.

There were also attempts to expand the range of American capability. These focused on the deployment of tactical nuclear weapons carried on Cruise and Pershing intermediate-range missiles. This deployment proved divisive in Western Europe, with particular concern about this deployment in West Germany. Both Reagan and Thatcher devoted considerable effort to winning support in Western Europe. The zero option was offered of no deployment if all Soviet intermediate-range missiles were removed from Europe, with Reagan keen on it as a first step for getting rid of all nuclear weapons. In contrast, Thatcher supported the measure only because she believed the Soviets would not agree: she wanted the American missiles deployed in order to counter Soviet conventional superiority, a reprise of the general strategy of nuclear deterrence, notably the NATO strategy.

Separately, there was an American commitment to the development of new space-mounted weaponry. The "Star Wars" program or Strategic Defense Initiative (SDI), outlined by Reagan in a speech on March 23, 1983, was designed to enable the United States to dominate space, using space-mounted weapons to destroy Soviet satellites and missiles. In 1986, albeit in a planned test, an American interceptor rocket fired from Guam hit a mock missile warhead dead-on. This test encouraged the Soviets to negotiate. It was not clear that the technology would ever work, in part because of the possible Soviet use of devices and techniques to confuse interceptor missiles. Indeed, Gorbachev was to support the Soviet army in claiming that the SDI could be countered.[23]

Nevertheless, the program was also a product of the financial, technological, and economic capabilities of the United States, and thus highlighted the contrast in each respect with the Soviet Union. The Soviets were not capable of matching the American effort, in part because they proved far less successful in developing electronics and computing and in applying them in challenging environments. Effective in heavy industry, although the many tanks produced had rather crude driving mechanisms by Western standards, the Soviet Union failed to match such advances in electronics. Moreover, the shift in weaponry from traditional engineering to electronics, alongside the development of control systems dependent on the latter, saw a clear correlation between technology, industrial capacity, and military capability. In the 1980s the Soviet Union fell behind notably.

In response to the Reaganite military buildup, the Soviet Union, with its anxious leadership fed intelligence reports about a hostile United States by the influential KGB, also adopted an aggressive pose. With the KGB providing totally inaccurate reports of American plans for a surprise nuclear first strike, the Soviets deployed more weaponry. In a huge drain on the resources of the already-strained Soviet economy, six *Typhoon*-class ballistic missile submarines entered Soviet service from 1980, as did their most impressive surface warships, including, in 1985, the launch of the *Admiral Kuznetsov*, their only conventional aircraft carrier. The *Typhoon*-class competed against the American *Ohio*-class, as the submarine evolved into an underwater capital ship as large as World War I *Dreadnought*-class battleships, and with a destructive capacity never seen before (or since) in any other type of warship.

More secretly, as an interview, published in the *New York Times* of February 25, 1998, with Kanatjan Alibekov, formerly an official in the program, revealed, the Soviet Union prepared anthrax and smallpox and plague virus cultures that would have been delivered by intercontinental ballistic missiles. This would truly have been anti-societal warfare. Nixon had abjured further American research and development of bacteriological and chemical weapons in 1969, a move that was not reciprocated by the Soviets. The Americans even began destroying their stores of these weapons. If the Soviets, into the early 1960s, had pursued bacteriological and chemical weapons as the "poor man's" alternative to tactical nuclear weapons, they still continued with the research when their tactical and strategic nuclear disadvantage had disappeared.[24]

The role of the KGB and the military helped ensure that foreign policy was scarcely under the control of Soviet diplomats. This role also contributed to the nature of Soviet foreign policy. The Ministry of Defense and its industrial ministry allies in the Council of Ministers, as well as the representatives of major industrial concerns in the Central Committee, directly and indirectly, were a major factor in Soviet foreign policy, which was reactive, hostile, and defensive, rather than attempting to seek international cooperation. Soviet diplomats were expected to respond to policies determined not by the Foreign Ministry but by a wide spectrum of office holders. Party unity was a key element. It reinforced the institutional conformity seen in the Soviet Union, a highly authoritarian state and society, with the general secretary of the Communist Party determining policy on what were presented as party lines. Conformity and cohesion ensured that information was prepared, presented, and analyzed accordingly.

There were alternative sources of information for foreign policy, notably regional institutes associated with the Soviet Academy of Sciences that were established for Latin America (1961), Africa (1962), Asia (1966), and North America (1967). The Maurice Thorez Institute of Foreign Languages was an outstanding and highly prestigious institution that provided superlative training in foreign languages and cultures for those destined to serve in the Soviet foreign service. This proved particularly valuable in Latin America. However, although the regional institutes had access to information about the outside world, their ability to influence the decision-making process was limited. Moreover, there was no Soviet equivalent to the often politicized and frequently open debate about policy options seen in the United States or to the alteration of power there (and in Britain, France, and West Germany) between the political parties. The Cold War thus saw a significant qualitative difference between the diplomatic systems in the two states, a difference that was very much in favor of the United States and its major allies.

There was a parallel difference in the intelligence services, with the KGB less ready to entertain debate than its Western counterparts. The Soviets proved better at descriptive intelligence gathering than at its analytic intelligence counterpart. The information from foreign agents was not synthesized in the manner seen in the West. Instead, the data entered depicted ever-increasing corroborations of what the "Center" expected to hear in relation to its orders to gather information on specific phenomena. Thus, the data served to buttress a priori assumptions. It was frequently not integrated and analyzed with respect to new analytic frameworks that the data might point to. This defect emerged glaringly during the NATO military exercise Able Archer in November 1983 when the Soviets overreacted, leading to the serious risk of war,[25] although this was not always the case. The KGB was divided into numerous competing sections or directorates, each of which was entrusted with specific responsibilities with respect to domestic and foreign intelligence gathering and operations. Although individual directorates could perform well, integrating the information provided (and thus integration already performed) by different directorates was often below the standard of American and British counterparts. The Soviets found it difficult to put together the empirical pieces. Whereas case officers who processed data could be accurate, it was often interpreted differently by their seniors. The same flaw could be seen with Western intelligence agencies, as in the assessment of developments in Cuba prior to the 1962 crisis.

COLD WAR TENSIONS, 1983–84

Cold War international tensions rose to a peak in 1983, with the deployment of Cruise and Pershing missiles in Western Europe exciting Soviet concern and anger, and the Soviets fearing attack under the cover of Able Archer. Moreover, the unrepentant Soviet shooting down on September 1, 1983, over Soviet airspace, of Korean Airlines flight 007, which they suspected of espionage, increased tension. Two hundred sixty-nine people, including an American congressman, were on the plane.

Decided on by NATO ministers on December 12, 1979, in response to the deployment of Soviet SS-20 intermediate range ballistic missiles in Eastern Europe, the Cruise and Pershing missiles arrived from November 1983, demonstrating the continued strength and effectiveness of the Western alliances. In addition, American rhetoric, notably Reagan's "evil empire" speech, which in some respects matched a long-standing Soviet pattern in rhetoric, rankled with the Soviet leaders. Moreover, the American invasion of the unstable, left-wing Caribbean island of Grenada in October 1983 accentuated Soviet concern about American actions and intentions. Yuri Andropov, the Soviet leader from 1982 to 1984, interpreted these actions to support his suspicions of the United States, and he suspended Soviet participation in the arms-control talks in Geneva. Andropov came out of *Gosbes* (State Security) and was a genuine ideologue. He believed in the inherent mendacity of Western imperialist leaders and society, and in imperialists' treachery and willingness to wage war against the Soviet Union. However, there was no precipitant to conflict, in part due to Soviet caution and in part because the Soviet Union could not afford war.

The situation was different in Central America and the Caribbean. In these areas, like the Soviet Union in Poland, the United States faced problems in a traditional sphere of influence and felt that it could and should act. This helped explain the American invasion of Grenada in 1983. This operation was motivated by concern about Grenada's leftward move, and the possibility that this would lead to a Cuban and Soviet military presence.

There was a tendency to see Grenada as another Cuba. The island was seized and the government changed.

A more sustained issue was posed by Nicaragua, where the Americans had intervened militarily against radicals in the 1920s. The left-wing Sandinistas, who gained control of Nicaragua, drew inspiration and support from Cuba, and also provided support for left-wing rebels in neighboring El Salvador. Concerned about the risk of instability throughout Central America and the wider regional challenge, and determined to mount a robust response, the Reagan administration applied economic, political, and military pressure on the Sandinistas, providing funds from 1981 to train and equip the Contras, a counterrevolutionary force that was based in neighboring Honduras. This was a limited war. Although the Contras helped to destabilize Nicaragua, inflicting considerable damage, they could not overthrow the Sandinistas. The Contra threat increased the bellicosity of the Sandinista state.

In El Salvador, where civil war had broken out in 1981, Reagan sought to protect, not overthrow, a government. Advisers, arms, including helicopter gunships, and massive funds were provided to help the right-wing junta resist the Farabundo Marti National Liberation Front (FMLN). However, the commitment of numerous American advisers was not followed by ground troops. This enabled the United States to define the struggle as low-intensity conflict; and thus, compatible with a definition of the Cold War centered on war avoidance. The conflict did not appear in this light to the population of El Salvador; they were caught between guerrillas and brutal counterinsurgency action that frequently took the form of terror. As with limited warfare, low-intensity conflict proved anything but for the civilians involved.

Tensions in the Middle East were accentuated by the crisis in Lebanon where Israel hit the Syrians hard in 1982 when invading southern Lebanon. In contrast to the situation in the 2010s, however, it proved possible to limit the wider spread of the struggle. The dispatch of a multinational Western force the same year, designed to try to bolster Lebanon's stability in the face of pressure from the Palestine Liberation Organization (PLO), competing militias, and Syria, was unwelcome to the Soviet Union, which provided support to both the PLO and Syria. In a demonstration of legitimation, the PLO was given diplomatic status, which was very much recognizing a combatant. Successful suicide attacks on the Western force in Beirut in 1983 led to its withdrawal. The United States did not take its commitment further. Conflict was limited.

CONCLUSIONS

The replacement of the dead Andropov, an adroit Cold War warrior who ran the country from a dialysis machine in his hospital room, by Brezhnev's protégé and former valet, the insipid and unimaginative Konstantin Chernenko, in February 1984, had scant effect in easing tension. Suspicious of Reagan, the Soviet leadership ignored his approaches for better relations. However, the practice of limited war had been sufficiently established to work. That is a bland summary of the continuation of tensions between West and East. At the same time, alongside ideological rivalry, there was nothing to approach the hatred reflected in the Iraq-Iran War of 1980–88. Like the Arab-Israeli wars, and those between India and Pakistan, this conflict brought a degree of religious animosity to the fore, one that in the Iranian case was accompanied by millenarian and existential drives.

The Cold War was subsequently to be held up as an instance of successful, or at least effective, deterrence[26] and of the practice of limited conflict, but that in part reflected, by the 1970s, both the consequences of the Sino-Soviet split and a degree of zeal in the Soviet Union that was less pronounced than that seen in some of the other conflicts of the latter half of the Cold War. There was stridency in the rhetoric, but it was not really a call to arms, and Western societies were certainly not mobilized to that end. The effectiveness of deterrence mechanisms was not tested to the extent that is often suggested or implied. Instead, the Cold War revealed a dependence on political stances, and this dependence was exemplified when changes in Soviet domestic policy led from 1985 first to the lessening of international Cold War tensions and subsequently to the political, economic, and ideological pressures that led to the collapse of the Soviet bloc in 1989 and of the Soviet Union in 1991.

NOTES

1. Roberts to Ernest Bevin, Foreign Secretary, NA FO 371/56763 fols. 37–38.

2. A. J. Rieber, "The Crack in the Plaster: Crisis in Romania and the Origins of the Cold War," *Journal of Modern History* 76 (2004): 62–106.

3. B. J. Bernstein, "The Uneasy Alliance: Roosevelt, Churchill, and the Atomic Bomb, 1940–1945," *Western Political Quarterly* 29 (1976): 214–16; J. Rose, "Winston Churchill and the Literary History of Politics," *Historically Speaking* 14, no. 5 (November 2013): 6.

4. *Foreign Relations of the United States: The Conferences at Malta and Yalta 1945* (Washington, DC: U.S. Government Printing Office, 1955), 450–51.

5. R. J. McMahon, *Dean Acheson and the Creation of an American World Order* (Washington, DC: Potomac Books, 2009).

6. NA FO 371/56763 fol. 26.

7. T. T. Hammond, *The Anatomy of Communist Takeovers* (New Haven, CT: Yale University Press, 1975).

8. M. P. Leffler, "The American Conception of National Security and the Beginnings of the Cold War, 1945–48," *American Historical Review* 89 (1984): 346–81.

9. C. S. Gray, "Mission Improbable, Fear, Culture, and Interest: Peacemaking 1943–49," in *The Making of Peace: Rulers, States, and the Aftermath of War*, ed. Williamson Murray and James Lacey (Cambridge: Cambridge University Press, 2009), 265–91.

10. *Foreign Relations of the United States 1947*, vol. 1 (Washington, DC: U.S. Government Printing Office, 1973), 80.

11. Alan Millett, *The War for Korea*, 2 vols. (Lawrence: University of Kansas Press, 2005–10). For a critical view of American policy, Bruce Cummings, *The Korean War: A*

History (New York: Random House, 2010), but see review in *Proceedings of the U.S. Naval Institute* 136, no. 12 (December 2010): 73–74.

12. Chancery to American Department, Foreign Office, August 2, 1950, NA FO 37/81655.

13. Ibid.

14. Rodric Braithwaite, *Armageddon and Paranoia: The Nuclear Confrontation* (London: Profile Books, 2017).

15. Nicholas Wheeler, "Interview with Robert Jervis," *International Relations* 28 (2014): 483.

16. Thomas Schelling, *Arms and Influence* (New Haven, CT: Yale University Press, 1966); Alexander George and Richard Smoke, *Deterrence in American Foreign Policy: Theory and Practice* (New York: Columbia University Press, 1974); Richard Ned Lebow, *Between Peace and War: The Nature of International Crisis* (Baltimore, MD: Johns Hopkins University Press, 1981); Robert Jervis, Richard Ned Lebow, and Janice Stein, *Psychology and Deterrence* (Baltimore, MD: Johns Hopkins University Press, 1984).

17. David Winkler, *Incidents at Sea: American Confrontation and Cooperation with Russia and China, 1945–2016* (Annapolis, MD: Naval Institute Press, 2017).

18. T. M. Nichols, "Carter and the Soviets: The Origins of the US Return to a Strategy of Confrontation," *Diplomacy and Statecraft* 13 (2002): 39–40.

19. John Gaddis, *The Cold War: A New History* (London: Penguin, 2005).

20. J. G. Wilson, "How Grand was Reagan's Strategy, 1976–1984?" *Diplomacy and Statecraft* 18 (2007): 773–803.

21. R. Higgs, "The Cold War Economy: Opportunity Costs, Ideology, and the Politics of Crisis," *Explorations in Economic History* 31 (1994).

22. S. Bronfeld, "Fighting Outnumbered: The Impact of the Yom Kippur War on the U.S. Army," *Journal of Military History* 71 (2007): 465–98; I. Trauschweizer, *The Cold War U.S. Army: Building Deterrence for Limited War* (Lawrence: University of Kansas Press, 2008).

23. Jonathan Haslam, *Russia's Cold War* (New Haven, CT: Yale University Press, 2011), 353–54.

24. D. E. Hoffman, *The Dead Hand: Reagan, Gorbachev and the Untold Story of the Cold War Arms Race* (London: Penguin, 2011).

25. Taylor Downing, *1983: Reagan, Andropov and a World on the Brink* (London: Little, Brown, 2018).

26. Ted Hopf, *Peripheral Visions: Deterrence Theory and American Foreign Policy in the Third World, 1965–1990* (Ann Arbor: University of Michigan Press, 1994).

Chapter Nine

War since the Cold War, 1990–

The war since 1990 that has attracted most controversy, including most attention in terms of the discussion of its causes, is the Iraq War of 2003. It brought together already pronounced debate over liberal interventionism with consideration, much of it highly critical, of the particular interests of great powers, notably the United States and its supposed drive for oil. At the same time, the slow-motion nature of the buildup to the war indicated that it was scarcely one that broke out because of a rapid response to a clash between armed forces.

Such clashes, however, occurred frequently, although generally without war. Thus, in 1996, soldiers from Greece and Turkey, which clashed over maritime boundaries, landed on an islet, Imia (Greek)/Kardak (Turkish), threatening war before an American-led mediation persuaded them to back down. In 2018, Greek soldiers on the nearby island of Ro fired warning shots at a Turkish coastguard helicopter, leading the helicopter to back down. This episode was more serious due to the marked deterioration of relations from 2017, with Greece blocking the extradition of Turkish officers accused of plotting against President Erdogan. In response, Turkey arrested two Greek border guards, leading Greece to reinforce its borders. Erdogan's emphasis on alleged external challenges served as a means to mobilize domestic support and suppress domestic criticism, and linked to his hostility to Kurds and his intervention in Syria.

In the background in the period from 1990 were fundamental changes in the international order, as well as significant developments in technology. Neither was responsible for the nature or extent of conflict in this period, which was particularly common in sub-Saharan Africa. However, each element, notably a shifting geopolitics, was important for the prospect of conflict between the great powers and, therefore, for the possibility of any of them engaging in war with the protégés of another. In the 1990s, the United States was able to attack Iraq and Serbia, both former protégés of the Soviet Union, because its weaker successor, Russia, was in no position to intervene and did not strongly seek to do so. By the 2010s, the situation was very different. Indeed, in 2018, having failed to act on behalf of Libya in 2011, Russia threatened the United States should it intervene against Syria, and did so convincingly.

GULF WAR, 1991

The most dramatic war involving a great power in the 1990s was the Iraq (Gulf) War of 1991. As with many wars, this began as a specific and quick operation, but that became a major struggle due to the intervention of others. The operation, scarcely limited as it involved the complete swallowing of one state by another, was the rapid and successful Iraqi conquest of Kuwait in 1990, which was driven by both need and opportunity. The need was for resources and prestige after the hugely expensive and lengthy Iran-Iraq war of 1980–88. The first was also launched by Saddam Hussein, the brutal dictator of Iraq, and reflected his propensity to turn to violent solutions and his grasping at apparent opportunities. In 1980, this occasion was provided by the 1979 Islamic Revolution in Iran, and his misguided sense that that revolution provided opportunities to tackle outstanding differences with Iran, to gain regional predominance, and to profit from dissidence in the Arab-inhabited southwest of Iran. In 1990, there was a similar mixture, the specific issue being long-standing Iraqi claims to Kuwait, which was presented as its nineteenth province, Iraq laying claim to inherit Ottoman pretensions. The invasion was launched under the cover of a "popular uprising" in Kuwait, with the "new government" calling for Iraqi help—a charade directed by Saddam Hussein. That Iraq had a large military was a crucial enabler, as well as providing the challenge, after the Iran-Iraq war, of finding it something to do.[1]

There was also a mistaken belief on the part of Saddam that the signaling by the American ambassador in Baghdad that America would not respond, the official policy confirmed shortly before the Iraqi invasion by the Department of State spokesman in Washington, would determine American actions. This was another iteration of mistaken readings of likely responses also seen with the North Korean invasion of South Korea in 1950 and the Argentine invasion of the Falklands in 1982: in response to the misreading of American and British steps, respectively. Militarized regimes may be particularly likely to misread the likely response of civilian-run states. In 1939–41, Germany and Japan underrated the persistence and resolve of democratic societies.

This misreading was an aspect of what can be seen as cultural and ideological elements, notably the misrepresentation and misunderstanding of others.

In 1990, the response was dramatic, not least because Iraq's move led to the assumption that it might be the first in a sequence of aggressive, even expansionist, steps, fulfilling Saddam's view of himself as the leader of the Arab world. Kuwait itself was not crucial, but its far larger neighbor, Saudi Arabia, was both the world's leading oil producer and a key American ally, one that reached to the Indian Ocean and the Red Sea. The importance of Saudi Arabia, in both strategic terms and those of oil production, were greatly increased due to the Islamic Revolution in Iran. The rapid deployment of substantial American forces to Saudi Arabia provided a means to engage with Iraq. Defensive forces, including the 82nd Airborne, were rapidly deployed under Operation Desert Shield, but the deployment of an American armored corps from Europe, a huge logistical exercise, for Operation Desert Sabre, took months. The American government initially sought the backing of the UN to produce both an international coalition (including quite large Egyptian and Syrian units) in support of its policy and a diplomatic means to persuade Iraq to back down. The failure to ensure the last led to the decision to attack. The deployment of American strength did not act as a deterrent sufficient to lead Iraq to back down. In the event, Iraq was defeated and driven from Kuwait in a highly destructive attack that was part of a limited war. An electronically enabled American conventional force utterly smashed Iraqi forces thanks to the use of comprehensive aerial surveillance and precision munitions. Once defeated, Saddam was not overthrown.

YUGOSLAVIA

This was very different to the outbreaks of conflict in the 1990s in former Yugoslavia and in the Caucasus, as the disintegration of Cold War alignments led there to the outbreak of atavistic levels of violence. In these, there was a degree of anti-societal violence stemming from the civilian-based nature of new military forces, but also from the extent to which ethnicity and religion were the identities created as the basis for, and goals of, conflict. The wars of the period in both areas are commonly discussed, as also with Somalia in the 1990s and Afghanistan in the 2000s, in terms of the intervention of outside powers. That approach downplays the role of local agency, not only in causing the original conflict, but also in affecting and encouraging the chance of outside intervention. This was certainly the case in the former Yugoslavia, Somalia, the Caucasus, and Afghanistan.

In each case, outside intervention totally reset the agenda, but it was hindered by the vitality of already-existing divisions on the ground. This process was to be further seen after the American-led invasion of Iraq in

2003. The wars in Yugoslavia were ultimately over the post–Cold War situation, but they were played out with reference to long-term tensions and anxieties. Religious animosity was one of the most significant, and played a major role in the Balkans, the Caucasus, and sub-Saharan Africa, notably, in the last, Ivory Coast, Sudan, and Ethiopia/Somalia. This led to talk of a war between civilizations, a concept focused on tensions between the West and Islam, although there was a marked tendency among Western proponents of this thesis to underrate divisions among the latter. Indeed, the latter were more destructive in terms of casualties than conflicts between Christendom and Islam.

A federal state held together earlier by its Communist dictator Josip Tito, Yugoslavia was divided between ethnic groups, most prominently Serbs and Croats, each of whom thought they were underrepresented and dominated by the other side. They sought independence for the areas they dominated and pursued the widest possible definition of the latter. Franjo Tudjman, the authoritarian president of Croatia from 1991 to 1999, used nationalism to provide both identity and rationale for his power, and the same was true of Serbia under Slobodan Milosěvić. In 1991, in the far north of the country, about 70,000 men out of a population of only two million Slovenes mobilized in order to resist attempts to prevent Slovene independence from Yugoslavia, and the Serb-dominated Yugoslav army did not push the issue to widespread conflict. It made a far greater effort in Croatia, which, unlike Slovenia, had a border with Serbia and also contained a large Serb minority. Thus, opportunity and apparent need encouraged a decision to fight that was not seen in the case of Slovenia.

This war spilled over into Bosnia, a part of Yugoslavia that was particularly ethnically mixed, with large Croat, Serb, and Bosnian Muslim populations, and that suffered from both Croat and Serbian expansion. Each of the communities in Bosnia formed an army, and the Bosnian Serbian and Bosnian Croat forces cooperated with the armies of Serbia and Croatia, pursuing both their own and joint objectives. The extent to which they pursued their own objectives made the conflict particularly complex. The conflicts in Yugoslavia were brutal, but also limited. War involved demonstration and negotiation, a politics by military means that was intensely political, and a mixture of sudden and brief brutality with truces, and convoluted strategies of diplomacy.

Western intervention to end the conflict was weakened by a combination of American reluctance, not least from the military leadership, and European weakness, but settlements were eventually imposed in Bosnia in 1995, and in Kosovo in 1999, at the expense of the expansionism and ethnic aggression of a Serbian regime that unsuccessfully looked for Russian sponsorship. Although the West played a major role, with 3,515 sorties flown (and 100 cruise missiles fired) in Operation Deliberate Force in 1995 (the first NATO combat mission), the ability of Serbia's opponents, especially the Croats, to organize military forces capable of opposing the Serbs in the field was important in preventing Serb victory.

This ability was seen in the autumn of 1995, when the Croats and the Bosnian Muslims, who had been brought together in large part by American pressure and support, were able to mount successful offensives against the Bosnian Serbs, the Croats overrunning first western Slavonia and then the Krajina. The attacking forces may have numbered 200,000 men. Combined with NATO air attack and diplomatic pressure, this pushed the Serbs into accepting the Dayton peace agreement that November. Outside commentators tended to focus on the NATO air attack as "the war," but the situation saw much more agency on the local scale.

The brutal slaughter of civilians by the Serbs (and, to a lesser extent, by their opponents) was an all-too-familiar feature of such conflicts in much of the modern world and reflected the extent to which ethnic groups were seen as the units of political strength, and thus as targets. In July 1995, the Bosnian Serbs murdered about 4,000 unarmed Muslim males in Srebrenica, which had been designated a safe zone by UN representatives whose peace-keeping force was too weak and too focused on self-preservation to prevent the massacre. What was termed ethnic cleansing—the expulsion, with considerable violence, often murderous and on a considerable scale, of members of an ethnic group—was more common. It was generally associated with the Serbs, but was also used by the Croats, for example in the Krajina, and, although the latter does not excuse Serb actions, it helps explain the paranoia that characterized their policy makers. Such action was very much war, however one-sided its nature. It also overlapped with "politics" within a number of states.

In turn, such action against civilians led to pressure on outside powers to adapt existing views on peacekeeping in order to adopt a proactive policy of peace enforcement focused on humanitarian goals. This overcame earlier hesitations about action, although not invariably so. For example, there was no such intervention in Rwanda in 1994. The choice of when to intervene and when not throws light on issues involved in the causes of war, not least the relationship between local struggles and international dimensions. In the specific case of Rwanda, the politics of learning lessons played a key role. Failure in an earlier intervention in Somalia in 1992–93, more particularly a perception of failure, led to American caution in 1994. Moreover, practicality and international power politics both played a role. Inland Rwanda did not lend itself to the amphibious power-projection capabilities the Americans had displayed in Iraq and Somalia. In addition, there were not the targets for air attack offered by Serbia. Rwanda was also seen as within the French sphere of influence and France was unwilling to act against genocide there, indeed seeing pressure for international intervention as meddling by the Anglosphere.

In order to suppress separatist demands, and to destroy support for the Kosovo Liberation Army, the Serbs, later in the decade, also used the tactics of "ethnic cleansing" in Kosovo, part of Serbia with a majority ethnic-Albanian and Muslim population. The Western response over Kosovo was coercive diplomacy, which in 1999 became a forceful humanitarian mission, Operation Allied Force. The resulting seventy-eight-day bombing and cruise-missile assault by American, British, and French forces in 1999 was less effective than Operation Deliberate Force had been in 1995, caused far less damage to the Serb military than was claimed, and did not immediately end the "ethnic cleansing," although it did help lead in 1999 to the Serb withdrawal and the Serb acceptance of a cease-fire, followed by the establishment by Britain, France, Germany, Italy, and the United States of a NATO peacekeeping force. The intervention led to much debate about the law on use of force. Thereafter, the continuing isolation of Serbia, in a form of economic and financial warfare, contributed to an erosion of support for Milosĕvić and his fall in the face of Serbian popular action in 2000.[2] In terms of the "start" of the conflict, it is unclear whether to put the emphasis on action by the Kosovars or the Serbs or NATO.

CIVIL WARS

The conflicts of the 1990s were discussed in terms of Mary Kaldor's concept of "New Wars"[3] and Rupert Smith's of "wars amongst the people."[4] Although both concepts are ripe for critique because they overemphasize the novelty of the situation, that decade, nevertheless, raised fundamental questions about governability within states and of the "world order." That looked toward the current and future situation, one exacerbated across much of the world by a rapid rise in population, and notably of young men. The economics of conflict within states attracted attention, and notably the overlap with drugs and with authority as a protection racket enforced by violence. That element was repeatedly seen in civil wars. It interacted with economic and political inequalities between and within communities. The violence is both top-down and bottom-up.[5]

There was also continuity from the conflicts in the Cold War, notably in parts of Africa, particularly Angola, and in Central America, especially El Salvador. The end of great-power support for one or the other side as the Cold War came to an end encouraged verdicts and/or settlements, as in Angola and El Salvador. However, these verdicts were not always permanent, as Afghanistan clearly showed. Religious and ethnic tensions, either separately or in combination, proved far more insistent as causes and supporters of conflict than the interventionist strategies that tend to engage attention.

Civil conflicts are a prime consequence of these tensions, as with the position of Muslims in Myanmar and Thailand. In the former, the exclusion from citizenship and state care of the Rohingyas, a Muslim minority, led in 2012 to the formation of the Arakan Rohingya Salvation Army, and from 2016 it initiated attacks on border guard and police posts using machetes and handheld explosives. In response, the army began a brutal, totally disproportionate repression of the Rohingyas as a whole, the vast majority of whom had nothing to do with the terrorism. Assisted by armed Rakhines, another local ethnic group, the army murdered, mutilated, and raped people and set houses on fire, causing over half a million people to flee. This was on the pattern of previous flights in 1978 and 1991–92 in which more than two hundred thousand fled. Moreover, it was on a pattern of the harsh treatment of other Myanmar minorities, such as the Shan, against which there were brutal campaigns of ethnic cleansing, as in 2009.

INTERVENTIONISM

Interventionist strategies have reflected a range of causes. Russian intervention in Chechnya was driven by factors that were very different to those of the United States in Bosnia. There was also a process of learning "lessons," frequently the wrong lessons. Thus, in 1992, notably but not only in the United States, confidence about the successful intervention against Iraq the previous year led to fresh intervention in Somalia, only for its apparent failure to lead not only to withdrawal from Somalia, but also to a decision not to intervene in Rwanda to limit genocide in 1994.

This process can be traced onward to the present: failed interventions in Afghanistan, Iraq, and Libya in 2001–11 were followed by great caution about intervening in Syria from 2012, and with Britain as well as the United States.[6] As yet, Russia has not encountered recent failure in such interventions, which means that no such lesson can be applied in debates there about action. Moreover, whereas Russian interventions recently, prior to that in Syria, were essentially by Russia alone, that has not been the case for the United States. Alliances and ad hoc coalitions create their own dynamics and tensions of causes and goals.

The American commitments of the period did not match what had become known as the Weinberger Doctrine, after Caspar Weinberger, secretary of defense from 1981 to 1987. This was articulated in the context of a dispute between him and Secretary of State Shultz over the disastrous 1983 American intervention in the Lebanon. Whereas Shultz wanted to use American force to back American diplomacy, Weinberger had pressed for commitments only in the event of predictable success and a clear exit strategy, and called for the use of overwhelming force. These priorities implied that the protection of the military took precedence over diplomatic goals and, in part, became the strategic object. Separately, the killing in a suicide bombing of 241 American Marines in Beirut was a key factor—because President Reagan did not respond—in the conviction that jihadism could succeed.

The varied causes of intervention throw more general light on those of war. Thus, for Russia, local domination played the key role in the decisions to act in Chechnya, Georgia, and Ukraine. Each was different, and in the first case there was fear that disorder in Chechnya might spread elsewhere in the north Caucasus and from there into Russia. In the northern Caucasus, Islamic independence movements were able to rely on considerable popular support. These movements tried to extend the breakup of the Soviet Union to Russia. Boris Yeltsin, the Russian leader, was unwilling to accept such separatism, not least because of the oil in the area. In 1994, he launched an invasion of Chechnya. The nature of Russian military policy—brutal and intransigent—encouraged resistance that the Russians were unable to crush and in 1996 they withdrew under a peace accord.

The renewed Russian attack on Chechnya in 1999–2000 was provoked by Chechen moves into neighboring Dagestan, and by explosions blamed on Chechen terrorists that may have been the work of the Russian secret police. The attack reflected the determination of Vladimir Putin, Yeltsin's prime minister and eventual successor as president, to assert his authority in the face of an insurrection that was not interested in limited success.[7] Moreover, he used a threat to consolidate his position, turning a situation into a war.

For Russia, there was no comparable fear of escalating separatism with its interventions in Georgia (2008) and Ukraine (2014), each former member states of the Soviet Union, but, instead, a more generalized anxiety that accepting democratic outcomes anywhere in neighboring states might lead to the spread of these aspirations. This powerful ideological drive was accompanied by the belief that compromise was weakness. That view was strongly expressed in the regime led by Putin, a regime that focused on the fact and use of power and did not regard democratic consent as a valid form of that power but, instead, as a threat to the regime. As an additional factor, the intervention in Ukraine, notably the seizure of Crimea, proved popular in Russia. It tied in with a sense of victimhood and with a pleasure in assertion.

In the case of the United States, similar attitudes had been seen in relations with Latin America over the previous century but were no longer central to policy making. By Western European standards, the United States was (and remains) overly ready to resort to force, and it is certainly clear, as Iraq exemplifies from 2003, that such intervention can dismally fail to achieve objectives. Nevertheless, the role of force in American governance and foreign policy, and of violence as a tool for policy, is different to the case of

Russia. Even more than during the Cold War, attempts to equate the two are deeply flawed and reveal a failure to understand the significance of the values that are important to strategic culture.

The decline of Cold War alignments increased volatility from the 1990s, although also providing opportunities, as in Angola, to end Cold War disputes. In the early 2000s, this volatility was taken further as a result of the suicide attacks launched by Osama bin Laden's *al-Qaeda* (The Base) terrorist movement on New York and Washington on September 11, 2001. These helped ensure that the American government took a more determined position in warfare in the early 2000s than had been the case in the Balkans in the late 1990s. The replacement of Bill Clinton by George W. Bush as president in early 2001 was also significant, although, ironically, Bush had opposed earlier "state-building."

The lessons of the 1990–91 Gulf War spurred American investment in the Revolution in Military Affairs (RMA) and the worldwide emulation of RMA capabilities. All around the world, militaries began to undertake transformation programs, all of them in preparation for future peer-to-peer warfare. In the event, 9/11 and the Global War on Terrorism presented a challenge to this vision of future warfare.[8]

"WAR ON TERROR"

At least initially after September 11, the Americans benefited from widespread international support in their self-proclaimed "Global War on Terrorism." There was much debate about the term; it was noted that to declare war on terrorism was nonsensical. Nonetheless, the term stuck because it gave American policy makers the flexibility to determine whom to target. The government felt it necessary in resisting terrorism, to strike back against it, not least as *al-Qaeda*'s use of suicide attacks meant that forward and preventative defense appeared essential and as there seemed no basis for a defense based on deterrence. This policy of striking back led to attacks, overt and covert, on what were identified as terrorist bases and supporters, which represented another stage in the movement toward action that had followed the end of the Cold War, not in a consistent pattern but frequently so.

In 2001, Russia lent diplomatic support to the American air offensive against the Taliban regime in Afghanistan, which had provided sanctuary for *al-Qaeda*. This backing occurred although this campaign, launched on October 7, entailed the establishment of American bases in Central Asian republics that had until 1991 been part of the Soviet Union, such as Uzbekistan under its autocratic president Islam Karimov. Russian support, however, was self-interested; the Russians sought American backing for their own campaign against Islamic opposition in Chechnya.

The American National Security Strategy issued in 2002 was both strategically and operationally ambitious. Pressing the need for preemptive strikes in response to what were seen as the dual threats of terrorist regimes and "rogue states" possessing or developing weapons of mass destruction—"America is now threatened less by conquering states than we are by failing ones"—the strategy sought to transform the global political order in order to lessen the chance of these threats developing. To that end, it proposed a universalist message.

In 2003, in Operation Iraqi Freedom, the Americans focused on Iraq—a definite and defiant target with regular armed forces—rather than on the more intangible struggle with terrorism, which challenged Western conventions of war-making. The attack was presented in terms of "drying up the swamp"—eliminating a state allegedly supporting terrorism, as well, more specifically, as destroying Iraq's supposed capability in weapons of mass destruction, particularly chemical and bacteriological warheads, although the reliability and use of intelligence about Iraq's weapons of mass destruction were subsequently to be discredited.[9]

ISRAEL

Specific wars interacted with more prolonged periods of lower-level conflict that occurred in a variety of forms. This can very much be seen in Israel's relations with its Arab neighbors. For example, the development in southern Lebanon of Hezbollah (the Party of Allah), an Islamic fundamentalist party and guerrilla organization that was armed by Iran and Syria, posed a major security threat to Israel. In response, and more particularly to a Hezbollah ambush in July 2006 of an Israeli unit patrolling the frontier, Israel decided to destroy Hezbollah's military power and capacity, especially its leadership, and its large and dangerous rocket arsenal. To that end, Israel blockaded Lebanon and launched a large-scale invasion of the south that month combined with extensive air attacks. This offensive, however, proved misconceived and poorly executed. A failure to appreciate Hezbollah's effectiveness had led to unrealistic expectations in Israel about success at the tactical, operational, and strategic levels.

Alongside this war came the more frequent pressure of lower-level conflict, a situation also seen in Israel's relations with the Hamas movement that is based in the Gaza Strip. As a further element, there was tension, sometimes violent, in Israeli relations with Palestinian Arabs on the West Bank. In each case, rhetoric, surveillance, policing, border controls, fortification, assassinations, and acts of violence played a significant role. The rhetoric was frequently that of war, but that does not settle the issue of definition.

CONCLUSION: CHANGING AMERICAN POLICY

Israel could not disengage from its neighbors, whereas the United States sought to do so from the Middle East in the 2010s. The Strategic Guidance document issued by the Pentagon in 2012 promised anew to defeat *al-Qaeda* and to counter the threat from unconventional weapons, but also drew back from interventionism and warned, instead, about China:

> In the aftermath of the wars in Iraq and Afghanistan, the United States will emphasize nonmilitary means and military-to-military cooperation to address instability and reduce the demand for significant U.S. force commitments to stability operations. . . . U.S. forces will no longer be sized to conduct large-scale, prolonged stability operations. . . . We will of necessity rebalance toward the Asia-Pacific region.

President Barack Obama had told the Australian Parliament on November 17, 2011, "As a Pacific nation, the United States will play a larger and long-term role in shaping this region and its future." This shift in priority was linked to a move from a two-war capability to a "win-spoil" plan entailing an ability to obtain victory in one regional war while thwarting the military plans of another adversary.

The shift was also driven by growing anxiety about China, anxiety that led to warlike talk. On June 2, 2018, speaking at the annual Shangri-La Dialogue, an Asian security summit in Singapore, James Mattis, the American defense secretary, warned of "larger consequences" if China continued its program of deploying offensive weapons to manmade islands in the South China Sea. In response, Chinese spokesmen declared that such comments were unacceptable, and that the weaponry was intended to protect against invasion. There was no meeting of minds.

NOTES

1. Kevin Woods, *The Mother of all Battles: Saddam Hussein's Strategic Plans for the Persian Gulf War* (Annapolis, MD: Naval Institute Press, 2008).
2. John Vasquez, "The Kosovo War: Causes and Justification," *International History Review* 24 (2002): 102–12.
3. Mary Kaldor, *New and Old Wars: Organized Violence in a Global Era*, 3rd ed. (Stanford, CA: Stanford University Press, 2012).
4. Rupert Smith, *The Utility of Force: The Art of War in the Modern World* (London: Allen Lane, 2005).
5. Paul Collier and Anne Hoeffler, "Greed and Grievance in Civil War," *Oxford Economic Papers* 56 (2004): 563–95; Frances Stewart, *Horizontal Inequalities and Conflict: Understanding Group Violence in Multiethnic Societies* (Basingstoke: Palgrave, 2008); David Keen, "The Economic Functions of Violence in Civil Wars," *Adelphi Papers* 38 (1998): 1–89 and "Greed and Grievance in Civil War," *International Affairs* 88 (2012): 757–77.
6. Adrian Johnson, ed., *Wars in Peace: British Military Operations since 1991* (London: RUSI, 2014).

7. Elena Pokalova, *Chechnya's Terrorist Network: The Evolution of Terrorism in Russia's North Caucasus* (Santa Barbara, CA: Praeger, 2015).

8. Theo Farrell, "NATO's Transformation Gaps: Transatlantic Differences and the War in Afghanistan," *Journal of Strategic Studies* 33 (2010): 671–99. For the opposition, Farrell, "Unbeatable: Social Resources, Military Adaptation, and the Afghan Taliban," *Texas National Security Review* 1, no. 3 (May 2018), https://tnsr.org/2018/05/unbeatable-social-resources-military-adaptation-and-the-afghan-taliban.

9. For a somewhat unconvincing attempt to offer a more sympathetic context, L. Freedman, "The Age of Liberal Wars," *Review of International Studies* 31 (2005): 93–107.

Chapter Ten

Into the Future

Optimism about a decline in the frequency of war has been repeatedly offered since the end of the Cold War. This optimism rests on a number of bases. These include claims about frequency that, however, beg many questions about how war, more particularly civil war, is defined, alongside trends related to such statistical measures. There are also arguments that the spread of nuclear weaponry acts as a deterrent and has made war less likely.[1] Each of these theses can be questioned for the present and past, and not least as to whether they are arguments that look toward the future. For example, the possession of nuclear weapons by those keen to use them would not act as a deterrent.

RESOURCE CRUNCHES

Moreover, whether optimism is compatible with resource issues against the dynamic of a rapidly rising world population is unclear. Indeed, looking to the past, environmental issues can be seen as playing a role in the causes of war. For example, climate change resulting in the shrinkage of pastures appears to have been a principal motor behind the flow of the Eurasian steppe nomads into the sedentary world.

Such issues prove particularly apposite in the case of tensions within states, being an important aspect of civil war as politics. For example, in South Sudan, since independence in 2012 there has been ethnic conflict between the Murle, Dinka, and Nuer tribes, with many thousands killed, in large part due to competition for land and cattle, and to raiding for children to use as slaves. Full-scale civil war broke out in 2013. By 2017, reports of atrocities, such as burnings alive, had become commonplace. The United Nations claimed that four million of the prewar population of twelve million had fled their homes. The key violence was between Dinkas and non-Dinkas, with the government and army run by the former, while non-Dinka armed groups unsuccessfully sought to prevent ethnic cleansing. Famine and disease followed.

Ethnic rivalry linked to resource competition played a role in internal conflict in many other African states, such as Ivory Coast between 2002 and 2011. In Togo, where Cnassingbé Eyadéma seized power in a military coup in 1967, he, or his son, Faure Gnassingbé, have been president ever since, basing their power on their minority Kabyé tribe, which has been given key posts in the expanded army. The Ewe and Mina tribes have been kept down. The notions that the world was becoming more peaceful, and that population increases were not necessarily creating conflict, looked less happy from the perspective of sub-Saharan Africa, for example the Central African Republic, Mali, and Nigeria, or, indeed, from that of Southwest Asia, than it does from Western Europe. Moreover, within Africa, the problems are spreading, as in Cameroon where the exploitation of western regions has led to a response that has caused fresh oppression.

Regional-ethnic conflicts are also found elsewhere. In Thailand, religion is a key problem with nearly seven thousand people killed in 2004–18 in a separatist insurrection in the separatist south. There, and in Myanmar, the government uses indiscriminate force as the first response. In Pakistan, Balochistan separatists angry with the exploitation of their region by the national government have met with a violent response.

Resource issues are also significant at the level of disputes between states. At the same time, it can be necessary to distinguish hyperbole and rhetoric from reality. For example, despite much talk of "water wars,"[2] in practice litigation or diplomacy have so far postponed, or even solved, most disputes. Dreams of total victory do not necessarily lead to war, although the emphasis on vindication, as opposed to hard-edged strategic thinking, that is characteristic of extremism, leaves few grounds for confidence in some areas, notably the Middle East.

THE COINCIDENCE OF CRISES

Writing against the background of the international crises of 2018 provides scant comfort, but also additional evidence that the causes of war are indeed twofold, first the bellicosity that encourages the resort to violence and, second, the difficulties of establishing limits in multilateral situations and relationships. In East Asia, this was true of China, Japan, North Korea, South Korea, and the United States, and in the Middle East of Iran, Israel, Russia, Saudi Arabia, Syria, Turkey, and the United States, not that that exhausted the list of participants. Indeed, from the perspective of the civil war in Yemen, there was a different list of Middle Eastern participants, including, for example, the United Arab Emirates. Major states found themselves committed to, or at least affected by, the actions of regional powers, and both could be affected by those of substate actors, such as Hezbollah and the Kurds in the Middle East, notably Syria and Iraq.

For all powers, there was also the question of how crises in one part of the world might interact with those elsewhere, and how the potential interaction would be assessed by other powers, both friendly and hostile. Thus, in 2018, the likely consequences for the Middle East of an American settlement with North Korea was an obvious point of interest, as was the prospect of conflict in the Middle East for relations in East Asia. Furthermore, America's capacity to act as a deterrent to Russia was affected by its capacity with regard to China, and vice versa. Related to capacity was the question of willingness, and the latter drew heavily on cultural assumptions.[3]

The potential relationship between domestic and international struggles was also significant. This was particularly so of the crisis in Syria, and, in that, followed the earlier crisis in Iraq. In contrast, there was no civil war in any of the states involved in crises in East Asia.

NEW WEAPONRY

The future trajectory of the causes of war is of course unclear. However, it is probable that the situation will change less than might be suggested. The technological rush of transformation in weapons systems does not currently imply a comparable alteration in the causation of war, although the development of more accurate atomic weaponry and the weaponization of chemical and bacteriological means give pause for thought.[4] So also with the possible impact of artificial intelligence (AI). A belief in "winnability," and of the dangers of not acting promptly, is important to the readiness to threaten, even risk, or possibly cause, war. At the same time, an aspect of this "winnability" today is the capacity to win the war with a small enough number of casualties to not lose the support of the public. Already significant in the West, this issue has become more prominent in Russia and, due to changes, including the impact of the "one child policy," also in China.

The crises of the 2010s so far have had much in parallel with those over the previous century, and that despite the major changes in technology. Nevertheless, alongside the hype, AI capability may change: from weapons able to carry out a specific task more effectively than humans, but still operating within human-controlled systems, to, on the other hand, fully autonomous weapons able to plan and solve problems. The latter may be two decades away, but procurement options looking to the future can affect planning in the shorter term.[5] Russia is openly investing heavily in AI technology.

RUSSIA AND CHINA

As ever, attitudes are crucial, and play a potent role in bellicose policy. Here, there is a subtle difference between Russia and China, although those exposed to their pressure may not see it. China makes warlike threats, notably toward Taiwan, and takes provocative and illegal steps in unilateral pursuit of its territorial claims, particularly by means of developing artificial island bases in the South China Sea. Russia, in contrast, has, under Vladimir Putin, openly attacked Georgia and Ukraine (at first covertly) and annexed Crimea, and the extent of its intimidatory action, notably in the Baltic, is greater than at any time since the 1980s. The Russian attack on Ukraine underlined the loss of deterrence by the latter caused by its "denuclearization" after the Cold War.

The Russian understanding of warfare as hybrid, along a spectrum from irregular to regular, informational and political to military, played a major role. The seizure from Ukraine of Crimea, by disguised Russian troops plus sympathetic locals, pushed Putin's approval rating to 86 percent, a rise of more than twenty points. Such action also helps him with his domestic image and popularity, and notably with his anti-Western supporters in the FSB, the intelligence service.

China and Russia clearly see the threat of force as crucial to their ability to pursue goals in their respective regions, the latter defined very much by them in terms of great power politics. This ability is further protected and enhanced by military technological change in the shape of the development of "anti-access/area denial" systems. Separately, both China and Russia are interested in power projection into distant areas, and notably so with the rapid development of Chinese naval power and overseas bases. The rhetoric is highly troubling, as are procurement policies focused on offensive weaponry.

THE VALUE OF COMMENTARY?

In each case, theory suggests that war is possible. In that of Russia, it is a declining power, notably in demographic and economic terms, that seeks to arrest its decline. In that of China, it is a rising power that both believes it deserves a larger role in the world and offers a different world order, including contrasting norms, whereas the United States, on the other hand, is an established power that feels threatened. This tension is referred to as the "Thucydides Trap," and discussed in terms of "power transition."[6] These factors have not yet caused war, but that is not a necessary guide to the future. The historical evidence for the theory of "power transition" is problematic. It has been argued that rising powers hardly ever go to war with dominant powers, and vice versa, that, linked to this, the few cases where they do have nothing to do with transition, and that transition occurs because of wars, rather than causes it. The "Thucydides Trap" in this view is a "total fiction, and a dangerous one."[7]

In practice, commentators, including theorists, often have relatively little to offer. In particular, in part due to the renewed pressures in the internet age to provide ready answers, there was a preference for clear-cut answers, sometimes expressed in paradigmatic forms, as opposed to an emphasis on the grit offered by events, contingencies, individuals, and the complexities of analyzing both the past and the present, let alone the future. For example, writing in the *Sunday Times* on April 22, 2018, Sir Lawrence Freedman confidently observed of Iran and Israel: "The foes will joust but know all-out conflict is too risky . . . they are not natural enemies . . . belligerent rhetoric should not be confused with a rush to war. A large-scale confrontation would carry enormous risks for both." Leaving aside the ironic counterpointing with an article in the same paper the same day entitled "Israel Braced for Attack by Iran as Shadow War Hits Boiling Point," as well as the series of articles to that end by the sister newspaper the *Times*, there are the simple points that "belligerent rhetoric" can very much help to frame responses to contingencies and, indeed, can create the latter.

THE MIDDLE EAST

In each case, there are also the uncertainties created by the political weaknesses on both sides. Each government is affected by disunity and domestic challenges, and this can increase the willingness to turn to external action. Moreover, on the part of Iran, there is an existential opposition to Israel that the Supreme Leader, Ayatollah Ali Khamenei, has compared to a cancer that needs to be cut out. He wants to unite the Shi'ite world behind what he sees as the liberation of Palestine, and the 120,000-strong Islamic Revolutionary Guard Corps is his tool. Iran has not only the dream of a Shi'ite empire, but also two governments. The Islamic Council is the central organ of a system of government that parallels that of "the government." An agreement with one is not necessarily a deal with the other. This helps make the Middle East highly unstable.

The commander of the Quds force, the foreign expeditionary army of the Revolutionary Guard Corps, Qassem Soleimani, is a key figure. To suggest that rhetoric should be ignored when considering individuals like Soleimani is mistaken. He prays alongside Khamenei and is close to him. While unlike

Iran, a theocracy, Israel is a democracy, it also has its own strong sense of religious destiny, indeed of unique religious destiny.

Soleimani is an instance of the role of "subgovernment agencies" in policy, both domestic and international. They undermine analyses of policy that focus on the state as a unit and, based on (an external analysis of) its interests, emphasize the pacific nature of its policy. For example, individual military units may have a role. So also with governmental elements. Thus, in China, the marine policy of Hainan Province "has direct repercussions for Chinese behavior at sea."[8] Commercial interests are also significant. The United Arab Emirates has pursued its geopolitical ambition in Djibouti through DP World, a major maritime firm largely owned by Dubai. In turn, this has led to a dispute through Djibouti, which increasingly depends on China.

Technology also offers the possibility of a "creep" or "slide" into larger-scale conflict. In June 2017, the United States shot down an Iranian-made armed Shaheed-129 drone over eastern Syria. Such action was warlike, but how far is it war? Much depends on the relationships between signaling and response. The character of the weaponry is also significant: an armed drone can be seen, and presented, as more threatening than a reconnaissance one.

Iran apparently seeks to create a continuous land route of control via Iraq and Syria to the Hezbollah region in Lebanon. This is a geopolitics in which religious sectarianism is important; Iran wants a cohesive Shi'ite bloc. Similarly, in Yemen, in the mid-2010s, the Houthis, a regional ethnic and religious grouping, were and are backed by Iran, which advanced the case of Shi'ite solidarity against the Sunni Saudis. Iranian intervention played a further role in encouraging Saudi attacks on the Houthis, attacks that, in 2015, brought the new king and his second son, Mohammed bin Salman, defense minister from 2015 and Crown Prince from 2017, valuable prestige within Saudi Arabia.

In Yemen, potent ethnic, sectarian, and regional differences, as well as large-scale intervention by several other powers, all challenged stability. The overthrow of one president, Ali Abdullah Saleh, in 2012, was followed by his return in alliance with the Houthis, whom he had previously long fought, albeit while his son had eventually developed links with them in order to undermine a powerful rival within the Yemeni military. Having driven the new president from the capital in 2015, the Houthis advanced into southern Yemen where his supporters had taken refuge, notably in the port of Aden. Neighboring Saudi Arabia intervened from 2015, as the lead power in an international alliance backed by the United States. Persistent Saudi air strikes, as well as the provision of arms and troops by the alliance, helped limit, but not reverse, Houthi success.

INTERNATIONAL DEVELOPMENTS

Developments may be discussed in terms of "rational" considerations: the equations of military and political profit and loss. However, that scarcely explains the wider context or the ability of the latter to determine, or at least affect, the operations of these considerations. The nature of conflict is also significant: all-out conflict may be unnecessary given the range of capabilities and means offered by types of hybrid warfare, by cyber-warfare, and by the use of proxies. So also with moves short of war, as with China's pressure on other states in the South and East China Seas, which is very much a pressure to conform. Although China is not apparently pursuing war, its policy is militarized and bellicose, transfers uncertainty to others, indeed imposes it on them, and is linked to the encouragement of nationalist sentiment within China, sentiment that is rising rapidly.[9] The situation with war and moves short of war, however, appears different if there is an attempt to control populated areas; that greatly increases the agency of locals and the issues and unpredictabilities involved in the responses to them. An engagement with "the human terrain" affects warfare and the risks involved, as Israel has found on the West Bank.[10] Moves short of full-scale war are a characteristic of Russian policy under Vladimir Putin, most recently in Ukraine and Syria.

It is certainly clear that the situation has become less stable as a result of the decline, certainly relative decline, of the American hegemon[11] and the failure to provide a system, stable or unstable, that may replace it. Deterrence has become more uncertain in this context and that uncertainty increases the risk of war. Moreover, the management of deterrence has become increasingly difficult, as seen in the Middle East, due to a lack of clarity, let alone certainty, over American policy. This is frequently blamed on Donald Trump, who became president in 2017, but, in practice, preceded him, and was particularly the case with Barack Obama's policies over Iraq and then Syria. As a result, Russia and Iran have become more active, while Israel and Saudi Arabia have struggled to engage American interest in the region, and notably did under Obama. It is unclear how far this situation reflects a response to a vacuum, in a form of quasi-automatic or formulaic system-operation, or how far reference to such a model involves a questionable attempt to provide a long-term analysis of much more contingent and unpredictable circumstances. Furthermore, the mechanistic connotations of the term vacuum may be unhelpful.

More significantly, there are fundamental ideological differences between a bloc that can be seen as the authoritarian states, one led by China and Russia and including Iran and Turkey, and the American-led opposing liberal bloc that includes Europe, India, and Japan. However riven by tensions, many historic, the first bloc is united in being primarily revisionist in international relations, whereas the latter bloc supports a rules-based international system. The very expression of their differences is an aspect of bellicosity. The Chinese government would not agree with the confidence that "there is simply no grand ideological alternative to a liberal international order,"[12] and is trying to create a bloc to follow its contrary lead.

In this context, contrasting beliefs about how to act most successfully can be significant, both contributing to bellicosity and, maybe, qualifying it. The sense that the Trump administration might act, and in a unilateral fashion, played a role, albeit an unclear one, in relations with North Korea in 2018.[13] More generally, the Russian belief under Putin that attacking gains the dominant position greatly affected NATO planning in the 2010s, but that also possibly increased the risks of conflict; NATO strengthening in Eastern Europe may well lead Russia to act.[14] NATO forces were deployed in Estonia, while both Lithuania and Sweden reintroduced conscription. Sweden also moved troops to its exposed Baltic island of Gottland, ratified a memorandum of understanding with NATO in 2016, and, in 2017, took part in NATO's annual Baltic training exercise. That year, the bellicose Vladimir Putin, in an interview with the Russian state-run news agency, Tass, warned:

> If Sweden joins NATO . . . we will consider that the infrastructure of the military bloc now approaches us from the Swedish side. We will interpret that as an additional threat and we will think about how to eliminate this threat.

This analysis focuses on "misperception and misjudgement" sparking "conflicts that both sides feel compelled to initiate."[15] That is a worthwhile position, but it downplays the very clear aggressive attitudes of a Russian government that, seeing democracy as a threat, inhabits a paranoid world in which the "color revolutions" in favor of democracy were presented as clearly designed to undermine Russia and its historical "rights." This approach is very much taken to the 2004 Orange and 2014 Maidan revolutions in Ukraine, the second of which overthrew the pro-Russian Viktor Yanukovych.[16] Russia's response to NATO's enlargement eastward has a paranoid character that rests on a long-standing narrative of the embattled motherland, a fusion of spiritual and secular interest.[17] Moreover, Putin's historicist approach to Russian identity, and his notions of geopolitics (drawing on those of Aleksander Dugin), are both predicated on the perpetual need to address threat and resist encirclement and containment.[18] In 2008, President George W. Bush publicly discussed the idea of Ukraine and Georgia joining NATO, only for Putin to warn that any NATO advance to its frontier would be taken as a direct threat to Russian security. France and Germany successfully opposed Bush's proposal, which they correctly saw as destabilizing. Later that year, Putin invaded Georgia's Russian-speaking provinces of South Ossetia and Abkhazia. On the part of Russia, there is clearly an ideological as well as a geopolitical analysis. At the same time, differing ideologies do not necessarily prevent the operations of an international order. [19]

So also with geopolitical factors. Areas can be seen in very different lights. For example, with "anti-access policies and weaponry," which, to a degree, are new terms for dominance of the sea, the relationship between the maritime and the land have changed. This change poses issues for the United States, since 1943 the leading naval power, which now has to determine whether it can retain the maritime character (notably free access to the sea) of areas increasingly dominated by the nearby continental power, China. [20] The technological, political-cultural, and policy factors involved are also pertinent for other areas where continental and oceanic interests and capabilities clash. That clash puts the focus on littoral areas, which are those in which the majority of the world's population lives.

In outlining future possibilities, or even clear trends, there is still the uncertainty presented by the lack of understanding of how best to deal with a combination of crises, such as affected Britain in 1940–41. Confronted by China, Russia, and Iran, the United States is arguably faced with a similar situation at present. This situation requires strategic acumen, [21] as well as resources and allies. Strategy involves making choices, [22] and, in doing so, also responding to the choices of others.

The military dimension can be seen with war-gaming and other plans. [23] The belief that war can be gamed using advanced Game Theory, however, underplays the extent to which in war there are no "rules" and always, instead, the surprise elements that cannot be constrained into the rules, no matter how smart the AI that is running the game. Moreover, the political dimension (both domestic and international) tends to be poorly covered in such games and planning, and understandably so as politics is difficult to reduce to a set of rules. Partly as a consequence, we are not too different from the situation with German planning prior to 1914, a military planning that devoted insufficient attention to political and economic dimensions.

CONCLUSION: "HYBRID" WARFARE: NOTHING NEW?

Looked at differently, the "hybrid," anti-societal warfare being developed by Russia, as in Ukraine, and also seen in Syria, represents a future for warfare that is intensely political. So also with the focus on affecting public opinion in other states. However, such warfare was indeed the usual pattern during the Cold War, in that both the Soviet Union and the United States, as well as other powers such as France and Britain, usually took part in conflicts without taking their involvement to the point of full-scale commitment or, indeed, hostilities. This was the involvement of special advisers, training missions, the provision of weaponry, the secret dispatch of special forces, and the persuasion of allies to take a role. Indeed, much of the international and military history of the late twentieth century can be understood in these terms. This was due both to the determination to avoid any escalation to a third world war and to American caution after the Vietnam War, and particularly so in contrast with earlier interventionism. The Soviet Union and its allies took a far more proactive stance, and notably in northeast and southern Africa. In each case, the Soviets operated by combining their advisers and weaponry with Cuban ground troops. Victory for Ethiopia over Somalia in the former, and for the Communist-backed MPLA over the Western-supported UNITA in Angola, indicated that this was a winning formula.

Somewhat differently, when preparing to attack Iraq in 2003 the American secretary of defense, Donald Rumsfeld, and other nonmilitary commentators, had been encouraged by the rapid American-backed overthrow of the Taliban regime in Afghanistan in 2001 to argue that airpower and special forces were the prime requirements, and that the large number of troops pressed for by the military was excessive. In the event, military pressure led to the allocation of sizable numbers for the attack on Iraq, although far fewer than in 1991. In the 2010s, there may well be a reversion to the earlier Cold War state of "hybrid" warfare, with comparable consequences for the causes of war, the two interacting in a complex process of cause and effect.

There are also the issues created by leaders who believe the myths that surround the "successful" military and its might, and the possibility that the military projects to the rest of the world. This tendency appears to be accentuated in states, such as Turkey, where senior officers are purged and the military is unable and unwilling to offer alternative advice, and notably to press for caution. This situation, that of the "loyal-incapability" army, increases the risk of military action to win reputation and/or to compensate for a loss of face. The current leadership of authoritarian regimes does not encourage optimism on this head about the supposed obsolescence of war.

NOTES

1. John Mueller, *The Remnants of War* (Ithaca, NY: Cornell University Press, 2004); Steven Pinker, *The Better Angels of our Nature: Why Violence Has Declined* (New York: Viking, 2011); Robert Jervis, "Force in Our Times," *International Relations* 25 (2011): 420.

2. For India, *Financial Times*, May 8, 2018.

3. Richard Ned Lebow, *A Cultural Theory of International Relations* (Cambridge: Cambridge University Press, 2008) and *Constructing Cause in International Relations* (Cambridge: Cambridge University Press, 2014).

4. Francis Gavin, *Nuclear Statecraft: History and Strategy in America's Atomic Age* (Ithaca, NY: Cornell University Press, 2012).

5. Paul Scharre, *Army of None: Autonomous Weapons and the Future of War* (New York: W. W. Norton, 2018).

6. David Rapkin and William Thompson, *Transition Scenarios: China and the United States in the Twenty-First Century* (Chicago: University of Chicago Press, 2013); but see Steve Chan, *China, the US and the Power Transition Theory: A Critique* (New York: Routledge, 2008); Graham Allison, *Destined for War: Can China and America Escape Thucydides' Trap?* (Boston: Houghton Mifflin, 2017). See, for criticism, Arthur Waldron, "There Is No Thucydides Trap," *SupChina*, June 12, 2017.

7. Richard Ned Lebow to Jeremy Black, e-mail, June 6, 2018; Richard Ned Lebow, *Why Nations Fight: Past and Future Motives for War* (New York: Cambridge University Press, 2010).

8. Ryan Martinson, "Panning for Gold: Assessing Chinese Maritime Strategy from Primary Sources," *Naval War College Review* 69, no. 3 (2016): 27.

9. Bill Hayton, *The South China Sea: The Struggle for Power in Asia* (New Haven, CT: Yale University Press, 2014).

10. Montgomery McFate and Janice Laurence, eds., *Social Science Goes to War: The Human Terrain System in Iraq and Afghanistan* (New York: Oxford University Press, 2015).

11. Simon Reich and Richard Ned Lebow, *Good-Bye Hegemony! Power and Influence in the Global System* (Princeton, NJ: Princeton University Press, 2014).

12. G. John Ikenberry, "The End of Liberal International Order?" *International Affairs* 94 (2018): 23.

13. For the argument, written before this crisis, for a degree of continuity, Hal Brands, *American Grand Strategy in the Age of Trump* (New York: Brookings Institution Press, 2018).

14. Edward McLellan, "Russia's Strategic Beliefs Today: the Risk of War in the Future," *Orbis* 61 (2017): 255–68.

15. Ibid., 268. For the Western failure to understand, Peter Roberts, "Designing Conceptual Failure in Warfare: The Misguided Path of the West," *RUSI Journal* 162, no. 1 (February–March 2017): 14–23.

16. Rajan Menon and Eugene Rumer, *Conflict in Ukraine: The Unwinding of the Post-Cold War Order* (Cambridge, MA: MIT Press, 2015).

17. Gregory Carleton, *Russia: The Story of War* (Cambridge, MA: Belknap Press, 2017).

18. Michel Eltchaninoff, *Inside the Mind of Vladimir Putin* (London: Hurst, 2018); Timothy Snyder, *The Road to Unfreedom: Russia, Europe and America* (London: Bodley Head, 2018).

19. Barry Buzan and George Lawson, *The Global Transformation: History, Modernity and the Making of International Relations* (Cambridge: Cambridge University Press, 2015); Andrew Phillips and J. C. Sharman, *International Order in Diversity: War, Trade and Rule in the Indian Ocean* (Cambridge: Cambridge University Press, 2015).

20. Peter Dutton, "A Maritime or Continental Order for Southeast Asia and the South China Sea?" *Naval War College Review* 69, no. 3 (2016): 12.

21. Arthur Waldron, "Could Four Simmering Global Crises Boil Over?" *Orbis* 60 (2016): 187.

22. J. Furman Daniel III, "Through a Glass, Darkly: Strategic Perspective(s) for an Uncertain World," *Orbis* 59 (2015): 293.

23. Edward Miller, *War Plan Orange: The U.S. Strategy to Defeat Japan, 1897–1945* (Annapolis, MD: Naval Institute Press, 1991); Albert Nofi, *To Train the Fleet for War: The U.S. Navy Fleet Problems, 1923–1940* (Newport, RI: Naval War College Press, 2010); John Lillard, *Playing War: Wargaming and U.S. Navy Preparations for World War II* (Lincoln: University of Nebraska Press, 2016).

Chapter Eleven

Conclusions

The willingness, for some need, to fight is the key element in causing fighting, at least in the form of enabling it. This willingness is shaped by bellicose drives that encourage and sustain war, seeing it as a form of "proving" oneself, one's group, and one's cause. These drives include the role of identity, honor, and reputation (including revenge) in the shaping of goals, and their commitment in the pursuit and sustaining of them. To treat the goals, and the mechanisms through which these are pursued, as the causes of war is to underplay the role of this commitment, which has been important across cultures and history. The typology offered in this book—of wars between cultures, wars within cultures, and civil wars—is helpful in this context of factors leading to commitment, but cannot itself explain variations in commitment.

Across history, it is instructive to turn to archaeological and artistic evidence for long-term cultural themes, many of which recur in very different settings, geographical, chronological, and with reference to types of war. For example, the remains of the Colima culture, which thrived in west Mexico from 100 BCE to 300 CE, indicate that its shamans (religious figures) communicated with the gods on behalf of the people and were also great warriors depicted with the heads of the defeated. The treatment of the defeated in this fashion was, and remains, a way to gain their potency and magic, and to demonstrate victory over the living and the dead. It can be seen in the desecration of bodies in modern conflict, notably in sub-Saharan Africa, with the castration of corpses being important. This process is not simply a sign of supposedly "primitive" cultures. For example, during World War II, the Germans not only slaughtered Jews and destroyed their culture, but also created a repository of the latter. It was as if Heinrich Himmler, the head of the SS, who was strongly committed to the occult, wanted to seize Jewish magic/ religion/belief in order to make that of the Nazis stronger.

The totemic character of conflict was also seen in the determination to hold onto the legacy of the past, of its honor and power. This has been seen across the cultures, from antiquity to the present, with military units eager to list past battle honors on their standards. Giovanni Panini's painting *Alexander the Great at the Tomb of Achilles* (c. 1718–19) depicted the episode in which Alexander, who believed he was descended from Achilles, allegedly ordered that the tomb of Achilles in Troad be opened so that he could pay tribute to the great warrior of the past, and thus acquire his magic.[1] There was a wider meaning in the popularity of images of Alexander's exploits; they came to validate Europeans' sense of their destiny in the world, as in Napoleon's approach to his conquest of Egypt in 1798 or Marzio di Colantonio's painting *Alexander the Great in His Conquest of Asia* (c. 1620). This was an aspect of the *traditio imperii*, the inheritance of Classical imperial power, that was so important in Christendom. This process was also very significant in other cultures, and notably in China when non-Chinese dynasties were established.

The validation of the past, or rather from the past, should not be presented in a mechanistic or instrumentalist fashion. In practice as well as ideology, the wars between religions, cultures, peoples, and states, the ideas of inherent conflict, and the very fact of conflict were lessons of history. As such, but also with a wider cultural value and potency, the wars were a matter of the trust between the generations, indeed the relationship between past, present, and future. That drive was important for societies reverential of the past and referential to it, which has been the case of the vast majority of societies throughout history. It was a drive that has cut across the idea of common humanity and the global citizenship of all.

War therefore rests on a deep identity in human society, that of conflict as an establishment and definition of nationhood and its boundaries, and as a link between the generations. This identity is clearly not universal. Indeed, the move against war in many societies since 1945, notably in Western Europe and Japan, is an important indication both of change and of the capacity for change, as well as of intra-generational rifts, even conflict, as with the United States during the Vietnam War. In addition, there are religious groups that, in seeking a ministry to all, can oppose the recourse to violence. This is certainly true of some modern branches of Christianity, such as Quakers, but also of modern tendencies in more mainstream churches.

In analytical terms, moreover, there is the thesis that war is becoming less common. This argument, however, is problematic due to the difficulty of assessing civil warfare. Furthermore, there are periods in the past that can appear as gaps between what turned out to be others of more sustained conflict: peace sometimes becomes simply the "interwar." A key problem of definition is offered by South America, where there have been no significant struggles between states since the end of the Gran Chaco War between Bolivia and Paraguay in 1935, but which is not therefore pacific.

The Gran Chaco War itself showed the role of choice. Long-standing competing territorial claims in the sparsely populated Chaco region that separated Bolivia and Paraguay had resulted in clashes in 1921 and 1927 and had nearly led to war already in 1928. That year, Colonel Rafael Franco of the Paraguayan army, acting without orders, seized and destroyed a Bolivian fort that had been built inside territory claimed by Paraguay, a key *casus belli*. He was sacked, as neither government yet wanted war, nor was ready to fight it, in particular lacking sufficient armaments. Nevertheless, Bolivia and Paraguay actively prepared for conflict, which was a more important background legacy.

In June 1932, there was a clash over a water source in the desolate Chaco, an important site as water was in short supply in the region, but one that could have been settled by negotiation or allowed to simmer. Instead, in a new departure, Bolivia escalated the dispute and turned to the attack. Daniel Salamanca, its determined president, wished to advance to the Paraguay River, which would provide a route to the sea and make landlocked Bolivia more of a regional power, as well as subjugating Paraguay. The route to the sea was a key geopolitical issue; it would enable Bolivia to export its minerals, especially tin, without being reliant on intermediaries who might refuse permission, charge high transit duties, and stop movements in time of war. Bolivia thus sought to compensate for the loss of its Pacific coastline to Chile in the War of the Pacific of 1879–83, providing a key instance of the role of the past in helping to cause war, albeit with a transfer of the sense of loss. In the event, Bolivia lost again and Paraguay was left in control of the Chaco.

Territorial disputes elsewhere in South America did not lead to full-scale conflict. For example, in 1932, a band of armed Peruvians took control of part of Amazonia then under Colombian rule. They had been organized by rubber and sugar entrepreneurs who had lost land in territory recently ceded to Colombia. Backed by local Peruvian military units, this move was then supported by the president, Luis Sánchez Cerro, a former colonel, who had seized power in 1930 and become a populist leader. In distant tropical rain forest, the sphere of operations was difficult for both powers. With Brazilian backing, Colombia recaptured most of the territory in 1933. The League of Nations played a role in negotiating a settlement on Colombia's terms, which also owed much to the assassination (for other reasons) of Sánchez Cerro. His successor, Oscar Benavides, did not have the same commitment to the issue.

Subsequently, territorial disputes in South America became less intense. In part, this absence of conflict was a product of an American hegemony that continued despite Cuba going Communist, albeit a hegemony that now looks increasingly precarious. Moreover, the lack of ideological division as a reason for any conflict between the states was important, for example in avoiding war between Argentina, Brazil, and Chile, which otherwise competed for primacy, with Argentina and Chile also having a border dispute.

Although Brazil sent forces to help the Allies in Italy in World War II, while Argentina attacked Britain by invading the Falklands in 1982, there have been no other conflicts between South American states and those elsewhere. Yet, that does not mean that there have been no wars in South America since 1935, because the use of force to advance political agendas has been important. This is true both of the military as the arm of the state and of attempts to overthrow the state or simply its government. The latter have come from the army, as in Argentina, Brazil, and Chile, and from groups that the army has been used to resist, as in Peru and Colombia. Insurrections have reflected social, political, regional, and criminal pressures. To present South America as if war is in decline may appear reasonable if the context is that of state-to-state conflict and/or the years since the Cold War, or if the focus is on statistics, but there is still the large-scale employment of force, as in Venezuela in recent years by the Maduro government in order to thwart opposition. This employment is a matter not only of the regular military, but also of paramilitaries and, indeed, hired thugs.

Breaking up rallies by demonstrators and employing violence, notably at election time, have also been seen over the last decade in other countries, particularly in Iran, Russia, and Thailand. In Turkey in 2017, there was the mass imprisonment of opponents of the government, notably since a failed coup in 2016, as well as the deployment of troops against any expression of Kurdish autonomy. These processes may not be marked as civil wars, and certainly do not approximate to, say, the American, Chinese, or Russian civil wars, but that does not mean that the basis for assessing changes in the quantification of war is clear, and particularly so if looking at civil war.

The use of force in disputes within states underlines the bellicosity of societies. It is difficult for many, indeed most, to accept the legitimacy of opposing views and of those who hold them gaining power, which is the essence of democracy; and, indeed, of religious toleration, freedom of opinion, and the role of law. These hostile attitudes are the essential cause of wars. So also with the lack of legitimating principles and practices in many states. The difficulty of ensuring legitimacy compromises consent, and governments are thereby exposed to rebellion.

At the same time, there have been changes across time. Whereas individual governments might rest on force, there are fewer states than in, say, 1400 whose coherence and territorial integrity essentially rest on the military prowess of the individual ruler and his ability to command or win over support through continued success. Thus, the functional and ideological value of war and victory have diminished in many cultures. The impression of power may well remain important in some societies to ensuring control, with conflict arising from the failure to manage the impression of power adequately, but power is not always the measure of legitimacy and consent.

The underlying lesson is of a continuity in the bellicose attitudes and behavior of the species as a social (as opposed to a solitary) animal, one that is not itself surprising. At another level, there are important variations, both geographically *and* temporally. These are between and within cultures,[2] which are themselves far from fixed or uniform categories. Indeed, what have been termed strategic cultures since the 1970s are themselves controverted and contested. The concept of a strategic culture, the assumptions and pressures arising from international relations and domestic contexts, has been very important. It is important to the setting of goals and the tasking that, in turn, helps determine military doctrine and force structure and that drives strategy. The mechanics of diplomatic processes, political consent, social context, and military means all vary geographically, each contributing to specific state strategic cultures.

Weaponry is sometimes presented as a "magic bullet" of systemic change, notably with atomic weaponry. Looking to the future, it is uncertain how far the deployment of large numbers of weapons employing artificial intelligence (AI), the chic new technology, will alter the situation. They definitely appear to offer greatly enhanced capabilities, notably in lethality and in limiting the risk to troops on one's own side. By 2018, at least four hundred partly independent weapons and robotic systems were under development in twelve states. The ability of weapons to learn, reason, and adapt would be a key enhancement.

What that may mean for the causes of war is unclear. It is uncertain how far existing, and new, international laws and ethics will be incorporated into their operation, or indeed can be. War as set by tactical issues thereby becomes a possibility, one that poses a new instance of the longer issue of local agents and agency. Control recurs as the key factor.[3]

It is also unclear what new weaponry may mean for the conduct of civil wars. More generally, issues of governability that are cultural as well as practical repeatedly come to the fore there, as, separately, do the willingness of states to intervene in such conflicts, a situation clearly seen in Syria in the 2010s. War shows the willingness of many to act with violence and the extent to which, as with the Vietnam War, total war in one area may be an aspect of limited war elsewhere, with each subject to different types of control and debate. In short, bellicosity and what can be regarded as "rational" strategies can overlap or may coincide. The relationship between them should not be seen in terms of polar opposites, but rather of the complexities and contingencies of particular circumstances. That approach downplays the

idea of general rules, in the sense of predictive models, but, instead, underlines the extent to which variety is the general rule.

In considering strategic culture, and its consequences, prestige is a key
element. Prestige is at once "irrational" and "rational," subjective and objective, the latter as far as the political purpose, ethos, and structures of states
were, are, and will be, concerned. Similarly, religious elements can be seen
as both irrational and rational. For example, they were important to Swedish
intervention in the Thirty Years' War in 1630 and were not incompatible
with Swedish interests as a whole.[4] Returning to prestige, European theorists
such as Francisco di Vitoria and Justus Lipsius argued that honor and glory
should not determine the recourse to war. However, in practice, wars repeatedly involve a struggle of will and for prestige, albeit with these being very
differently understood in specific political, social, and cultural contexts.
These elements are particularly significant to the more general process by
which contrasting priorities were assessed and policy cohered.

In most cases, when considering the causes of war historically (and in the
present), it is possible to accumulate reasons for war, but without that accumulation necessarily explaining the drive or establishing priorities. These
difficulties are apparent whether the prime perspective is that of the individual state or of the international system. This is possibly particularly so when
states were involved already in another conflict. Thus, from 1677, the Ottoman empire was at war with Russia, in part due to Russian expansionism into
Ukraine, and to the Ottomans' determination to protect their position there
and that of their protégé, the Khan of the Crimean Tatars.

There was scant reason for that war to end, but a revolt in those parts of
Hungary ruled by the Habsburgs encouraged Kara Mustafa, the grand vizier,
to negotiate peace with Russia in 1681 and to begin aiding the rebels. This
can be presented in prudential terms and by focusing on the Ottoman leadership. The rebels under Imre Thököly might serve as the basis for a Hungarian
client state able to protect Ottoman borders and would certainly have weakened Austria. The views of Kara Mustafa and Sultan Mehmed IV were clearly important. The former believed that prestige gained from victory would
strengthen his position, and he commanded the army that besieged Vienna in
1683. The Ottomans also believed Austria to be weak and were encouraged
by the latter's enemy, France.

At the same time, there were also broader social circumstances explaining
war, notably the synergetical relationship between Ottoman military activity
and the interests of the provincial elite. Cultural factors included the focus of
opposition to infidels on the Catholic center of Vienna, which gave a com-
manding and meritorious target for Ottoman bellicosity. Opportunity and
ideology also helped explain the international support Austria received, both
in opposing the siege in 1683, and, after victory then, in forming a Holy
League in 1684.

Again, the theme of variety in the factors that can be cited in this and other instances illuminates the misleading tendency toward a primitivization, and thus simplification, of non-Western warfare.[5] In fact, there were processes of debate there as well as in Western cases. As a result, patterns and developments in the causes of war appear clear only if the complex nature of conflict in the past is not grasped. That point is also germane in response to ready statistical measures of war in which causes are allocated with reference to a clear categorization of wars and the related typology. Instead, it is the very ability of humans and human society to encompass differing values and drives that helps give complexity to the causes of war. That may not be the conclusion some readers want. They may prefer to note the emphases here on ideologies, including religion, as the cause, although not occasion, of bellicosity and the crucial willingness to kill and to support killing.

This element, however, does not explain the "occasions of war," in other words why there is an outbreak of hostilities at particular moments. Again, there is no general pattern. The most significant factor is the determination to fight on the part of the leadership of at least one of the powers involved. This determination can generally be described and explained without reference to the other power. While it may appear logical to argue that war requires willingness by both sides, in practice the decision by one power to attack is usually the cause. The views on international relations of the leadership, the apparent exigencies of domestic circumstances understood in a political perspective, and the psychological drives of the leaders or leader, are the key elements.

NOTES

1. Guy Hedreen, "The Cult of Achilles in the Euxine," *Hesperia* 60 (1991): 313–30.

2. Patrick Porter, *The Global Village Myth: Distance, War, and the Limits of Power* (Washington, DC: Georgetown University Press, 2015).

3. Kenneth Payne, *Strategy, Evolution, and War: From Apes to Artificial Intelligence* (Washington, DC: Georgetown University Press, 2018).

4. Erik Ringmar, *Identity, Interest and Action: A Cultural Explanation of Sweden's Intervention in the Thirty Years War* (Cambridge: Cambridge University Press, 1996).

5. Patrick Porter, *Military Orientalism: Eastern War Through Western Eyes* (London: Hurst, 2009).

Selected Further Reading

Afflerbach, Holger, and David Stevenson, eds. *An Improbable War? The Outbreak of World War I and European Political Culture before 1914* (Oxford: Berghahn Books, 2007).

Albert, Mathias. *A Theory of World Politics* (Cambridge: Cambridge University Press, 2016).

Allison, Graham. *Destined for War: Can America and China Escape Thucydides's Trap?* (Boston: Houghton Mifflin, 2017).

Barfield, Thomas. *The Perilous Frontier: Nomadic Empires and China, 221 BC to AD 1757* (Cambridge, MA: Blackwell, 1989).

Black, Jeremy. *Beyond the Military Revolution: Warfare in the Seventeenth-Century World* (Basingstoke: Palgrave, 2011).

———. *Rethinking Military History* (London: Palgrave, 2004).

Blainey, Geoffrey. *The Causes of War* (New York: Free Press, 1973).

Copeland, Dale. *The Origins of Major War* (Ithaca, NY: Cornell University Press, 2000).

Gat, Azar. *War in Human Civilization* (Oxford: Oxford University Press, 2006).

Glete, Jan. *War and the State in Early Modern Europe* (London: Routledge, 2001).

Goertz, Gary, and Jack Levy, eds. *Explaining War and Peace: Case Studies and Necessary Condition Counterfactuals* (New York: Routledge, 2007).

Gommans, Jos. *Mughal Warfare* (London: Routledge, 2002).

Hunt, Peter. *War, Peace, and Alliance in Demosthenes' Athens* (Cambridge: Cambridge University Press, 2010).

Lebow, Richard Ned. *Why Nations Fight: Past and Future Motives for War* (New York: Cambridge University Press, 2010).

Levy, Jack, and William Thompson. *Causes of War* (Chichester: Wiley-Blackwell, 2010).

Lorge, Peter. *The Asian Military Revolution* (Cambridge: Cambridge University Press, 2008).

Murphey, Rhoads. *Ottoman Warfare* (London: Routledge, 2000).

Nish, Ian. *The Origins of the Russo-Japanese War* (Harlow: Longman, 1996).

Payne, Kenneth. *Strategy, Evolution, and War. From Apes to Artificial Intelligence* (Washington, DC: Georgetown University Press, 2018).

Perdue, Peter. *China Marches West: The Qing Conquest of Central Eurasia* (Cambridge, MA: Harvard University Press, 2005).

Porter, Patrick. *The Global Village Myth: Distance, War, and the Limits of Power* (Washington, DC: Georgetown University Press, 2015).

———. *Military Orientalism: Eastern War Through Western Eyes* (London: Hurst, 2009).

Ringmar, Erik. *Identity, Interest and Action: A Cultural Explanation of Sweden's Intervention in the Thirty Years War* (Cambridge: Cambridge University Press, 1996).

Schmitt, Carl. *Writings on War* (Cambridge: Polity, 2011).

Smith, Iain. *The Origins of the South African War, 1899–1902* (Harlow: Longman, 1996).

Smith, Joseph. *The Spanish-American War: Conflict in the Caribbean and the Pacific, 1895–1902* (Harlow: Longman, 1994).

Stevenson, David. *Arms Races in International Politics: From Nineteenth to Twenty-First Century* (Oxford: Oxford University Press, 2016).

———. *The Outbreak of the First World War: 1914 in Perspective* (Basingstoke: Macmillan, 1997).

Vasquez, John. *Contagion and War: Lessons from the First World War* (Cambridge: Cambridge University Press, 2018).

———. *The War Puzzle Revisited* (Cambridge: Cambridge University Press, 2009).

Wright, Quincy. *The Causes of War and the Conditions of Peace* (London: Longmans, 1935).

Index

Germany: Army Bill (1913), 112; Austria alliance with, 105–106, 108; Belgium invaded by, 117–118; declaration of war by, 117; Mexico connection to, 128; militarism unifying, 106; mobilization of, 115; navy ambitions of, 110–112; submarine warfare by, 127–128; Wars of German Unification, 103, 119, 121; World War I preparations for France, 109–110; World War I preparations for Russia, 108–109. *See also* Nazi Germany
Gibbon, Edward, 62, 65
globalization, 9
Global War on Terrorism, 1; interventionism and, 205; Iraq War (2003) and, 206; against al-Qaeda, 205, 207
Gran Chaco War, 223
Great Northern War (1700–21), 51–52
Greece: Britain military aid to, 170; Turkey conflict with, 197; World War I entrance of, 126; in World War II, 154
Greek Civil War, 175, 176
Grenada, 193–194
gunpowder empires, 26

Hegelian-Marxist assumptions, 59–60
heroism, 42–43
Himmler, Heinrich, 221–222
History of the Peloponnesian War (Thucydides), 11, 20
History of the Reign of Charles V (Robertson), 64–65
Hitler, 5; declaration of war by, 163; Lloyd George and, 144–145; Munich Agreement (1938) broken by, 145, 147; Mussolini similarities with, 142, 153; racial ideology toward Jews, 138, 155–156; Stalin in relation to, 147–148. *See also* Nazi Germany
hooliganism, 2
Horn of Africa, 34
Hungary: Nazi Germany alliance with, 133, 157–158; Ottoman Empire and, 226
hunting origins, 15–16
Hussein, Saddam, 198–199
hybrid states, 56–57

hybrid warfare: assassination as, 7; background on, 217–218; World War II and, 4

identity: patriotism and, 41; war for, 222
ideology: of expansionism, 105; Hitler's racial, 138, 155–156; of masculinity, 104; role in religious conflict, 56; of total war, 127
imperialism: Alexander the Great influencing, 222; Christianity and, 88; state government advancing, 89–90; territorial ambitions and, 82–83; transformative conflicts caused by, 69; wars of, 82–90
imperial overstretch, 54
Incidents at Sea agreement (INCSEA), 181
India: Britain at war with, 83; civil war in, 65
industrialization, 119
international crises, 210–211
international developments, 215–217
international relations (IR): alliances for, 37; bellicosity and, 11, 12, 215–216; developments in, 215–217; Jiang Jieshi on, 159; nuclear confrontation and theory of, 181; strategic culture and, 225. *See also* alliances
interventionism: globalization linked to, 9; Russia and, 203, 204; United States and, 203, 204–205
interventionist warfare, 12
interwar, 222–223
intracultural conflicts: Anglo-Dutch War as, 39; regarding dynasticism, 39–42, 44–45; heroism and, 42–43; Manchu conquest as, 36–37; Ottoman-Mamluk wars as, 23, 35–36; psychology impacting, 43–44; in sixteenth century Europe, 36, 37, 37–38; Sunni and Shi'a divide as, 36, 214; Thirty Years' War (1618–48) as, 38–39
IR. *See* international relations
Iran: Islamic Revolution in, 198, 199; Israel foe of, 213, 213–214; Trump on, 3
Iraq War (1991), 198–199
Iraq War (2003): Global War on Terrorism and, 206; hybrid warfare regarding,

About the Author

Jeremy Black graduated from Cambridge University with a starred First and did graduate work at Oxford University before teaching at the University of Durham and then at the University of Exeter, where he is professor of history. He has held visiting chairs at the United States Military Academy at West Point, Texas Christian University, and Stillman College. He is a senior fellow of the Foreign Policy Research Institute. Black received the Samuel Eliot Morison Prize from the Society for Military History in 2008. His recent books include *Air Power: A Global History*; *War and Technology*; *Naval Power: A History of Warfare and the Sea from 1500 Onwards*; *Rethinking World War Two: The Conflict and Its Legacy*; and *Fortifications and Siegecraft: Defense and Attack through the Ages*.

Lightning Source UK Ltd.
Milton Keynes UK
UKHW041854281020
372396UK00001B/23